Structural Analysis and Design to Prevent Disproportionate Collapse

Structural Analysis and Design to Prevent Disproportionate Collapse

Feng Fu

City University, London, UK

CRC Press
Taylor & Francis Group
Boca Raton   London   New York

CRC Press is an imprint of the
Taylor & Francis Group, an **Informa** business

A SPON PRESS BOOK

CRC Press
Taylor & Francis Group
6000 Broken Sound Parkway NW, Suite 300
Boca Raton, FL 33487-2742

Printed on acid-free paper
Version Date: 20160122

International Standard Book Number-13: 978-1-4987-0679-7 (Hardback)

Visit the Taylor & Francis Web site at
http://www.taylorandfrancis.com

and the CRC Press Web site at
http://www.crcpress.com

Contents

Preface

The first time I was involved with disproportionate collapse prevention design was in 2007. That was the first time I encountered design guidance for progressive collapse, such as those from the General Services Administration (GSA) and the Department of Defense (DOD). One of the difficulties I noticed at that time was that there was no software available to accurately perform progressive analysis for tall buildings or other complex structures, such as long-span space structures. Therefore, working together with my colleagues and using the general-purpose program Abaqus®,* I developed the progressive collapse analysis method based on the alternative path method. This method was found to be effective and has since been used for several other projects.

After that, I moved into academia and continued my research in progressive collapse analysis of different types of structures, such as tall buildings, bridges, and space structures, under extreme loading, such as fire and blast. I accumulated extensive experience in the area of structural design to prevent disproportionate collapse.

In the meantime, I also noticed that most design engineers lack knowledge of the theories and modelling techniques in progressive collapse analysis and design of complex structures. For students, there is also a large knowledge gap, as there are few textbooks available that cover the design and analysis of a structure to prevent disproportionate collapse.

The motivation for this book was to provide engineers with an understanding of the disproportionate collapse problems for different types of structures under different loading regimes, effective methods to model and analyze these types of structures using conventional commercial software such as Abaqus®, ETABS, and SAP2000, and the theories and design principles that underpin the relevant analysis. Another objective of this book is to provide civil engineer students with detailed knowledge in design and progressive collapse analysis of complex structures. Therefore, this book has been written not only to serve college and university students, but also as a reference book for practicing engineers.

---

* Abaqus® is a registered trademark of Dassault Systemes S.E. and its affiliates.

This book covers almost all the types of structures that an engineer may face, such as multistorey buildings, space structures, and bridges. It also covers effective methods to prevent the progressive collapse of each type of structure. Different loading regimes, such as fire and blast, which can trigger the progressive collapse of the structures, are also covered.

Another feature of this book is that it demonstrates three-dimensional (3D) modelling techniques to perform progressive collapse analysis for different types of structures through the examples, which replicate real collapse incidents and prestigious construction projects around the world, such as progressive collapse analysis of the Twin Towers, structural fire analysis of World Trade Center 7, blast analysis of the Murrah Federal Building, and progressive collapse analysis of the Millau Viaduct. This is to help engineers understand the effective way to analyze the structures to prevent their progressive collapse.

Feng Fu

Acknowledgements

I would like to express my gratitude to Dassault Systems and its subsidiaries, Computer and Structures, Inc. and Autodesk, Inc., for giving me permission to use images of their product.

I would also like to thank BSI Group in the UK, Crown Copyright in the UK, and the National Institute of Building Sciences in the United States for allowing me to reproduce some of the tables and charts from their design guidance.

I also thank the National Institute of Standards and Technology, Technology Administration, U.S. Department of Commerce, for allowing me to reproduce some of the images from its reports.

I want to express my gratitude to Forsters & Partners for providing some of the architectural drawings of the Millau Bridge projects I demonstrate in this book.

I am thankful to all reviewers who offered comments. Special thanks to Tony Moore and Kathryn Everett from Taylor & Francis Group, LLC for their assistance in the preparation of this book.

Thanks to my family, especially my father, Changbin Fu, my mother, Shuzhen Chen, and my wife, Yan Tan, for their support in finishing this book.

About the Author

Dr. Feng Fu earned his PhD from the University of Leeds and MBA from the University of Manchester. He is a chartered structural engineer, member of the Institution of Structural Engineers, the Institution of Civil Engineers, and the American Society of Civil Engineering. He is also a committee member of Disproportionate Collapse Mitigation of Building Structure Standards and Blast Protection of Building Standard, American Society of Civil Engineers. He is on the editorial board of *Advances in Computational Design*, an international journal.

He is currently a lecturer in structural engineering at City University London, following his work at the University of Bradford in the same position. Prior to that, he worked for several world-leading consultancy companies, such as WSP Group, Waterman Group, and Beijing Institute of Architectural Design and Research. During his industrial practice, he was involved in the design of extensive prestigious construction projects worldwide, and has been working with several world-leading architects. He also gained extensive experience in designing buildings under extreme loadings, such as blast and fire, and knowledge in designing buildings to prevent progressive collapse.

Dr. Fu has extensive research experience in the areas of progressive collapse, structural fire analysis, and blast analysis of tall buildings and long-span structures. He conducted his research in the area of tensegrity structures and steel and composite structures. He specialized in advanced numerical modelling and developed several modelling programs using different languages. He is currently supervising four PhD students. Dr. Fu has published more than 30 technical papers and 1 textbook. He is also a peer reviewer for 18 international journals and 4 books of internationally renowned publishers.

Introduction

1.1 Aims and Scope

Disproportionate collapse or progressive collapse first attracted the attention of engineers when in 1968, Ronan Point, a 22-storey apartment building in London, collapsed (Ministry of Housing and Local Government, 1968). The events of September 11, 2001 (NIST NCSTAR, 2005), which caused the collapse of the Twin Towers in New York, are another milestone in the research and new design measures to resist progressive collapse of buildings. The incident caused several researchers to focus on the causes of progressive collapse in building structures, seeking the establishment of rational methods for the assessment and enhancement of structural robustness under extreme events. The 9/11 attack also caused increasing enforcement of new design guidance to prevent progressive collapse of different types of structures.

Since the collapse of Ronan Point, detailed structural design guidance for preventing progressive collapse has been developed in both the UK and United States, such as the British Building Regulations (HM Government, 2013) and BS 5950 (BSI, 2001) in the UK and guidance from the Department of Defense (DOD, 2009) and the General Services Administration (GSA, 2003) in the United States.

As a design engineer, it is imperative to guarantee that sufficient measures in the design process of a structure have been made to prevent the progressive collapse of the structure. An engineer should also have the capacity to analyse the progressive collapse potential of a structure using appropriate procedures and analysis software.

Therefore, this textbook is designed to help design engineers or structural engineering students fully understand relevant design guidance and analysis procedures. As progressive collapse analysis is a distinctive and complicated procedure, it normally requires an ability to use a modern commercial finite-element package. This book features a detailed introduction to the use of finite-element programs such as Abaqus®, SAP2000, and ETABS in this type of analysis. In addition, case studies based on various types of structures, such as multistorey buildings, long-span space structures, and bridges, are provided to demonstrate failure mechanisms and effective mitigation methods in design practice.

Chapter 1 introduces the definitions of *disproportionate collapse* and *progressive collapse*, followed by the introduction of robustness and relevant design guidance around the world.

Chapter 2 introduces several collapse incidents of multistorey buildings. It specifically focuses on the reason and mechanism of the Twin Towers collapse. In addition, relevant design and analysis methods to prevent the disproportionate collapse of multistorey buildings are introduced. At the end of the chapter, a modelling example of the progressive collapse analysis of the Twin Towers is presented using a general-purpose program, Abaqus®.

Chapter 3 introduces several collapse incidents of long-span structures, including the collapse at Charles de Gaulle Airport and other space structures. The reason and collapse mechanism of the space structure are also introduced. At the end of the chapter, a modelling example of the progressive collapse analysis of a double-layer grid is presented using Abaqus®.

Chapter 4 introduces several collapse incidents of bridges due to different reasons, such as a lorry strike and an earthquake. Then, these triggering events that caused the collapse are discussed in detail, and relevant design and analysis methods to prevent disproportionate collapse are introduced. At the end, a modelling example of progressive collapse analysis of the Millau Bridge is presented using Abaqus®.

Chapter 5 covers the basic knowledge of fire. Then, the incidents of the collapse of buildings due to fire are introduced, and relevant design and analysis methods for structural fire design and prevention of disproportionate collapse are introduced. At the end of the chapter, a modelling example of the structural fire analysis of World Trade Center 7 is presented using Abaqus®.

Chapter 6 introduces incidents of the collapse of buildings due to blast loading. It gives the basic knowledge of blast loading, so readers can understand the blast loading and the response of building components, and how to classify the level of damage. Then, relevant design and analysis methods for preventing disproportionate collapse are introduced. At the end of the chapter, a modelling example of the structural blast analysis of the Alfred P. Murrah Federal Building is presented using Abaqus®.

1.2 Definition of Progressive Collapse or Disproportionate Collapse

So far, the terms *progressive collapse* and *disproportionate collapse* have been found in many technical papers, and there are different definitions for them. This makes it difficult for engineers to understand the clear difference between them.

Here the definitions for these two terms that are considered the most accurate by the author are presented. *Progressive collapse* is defined as "the spread of an initial local failure from element to element, eventually resulting in the collapse of an entire structure or a disproportionately large part of it" (ASCE, 2005). *Disproportionate collapse* is defined as "a collapse [that] results from small damage or a minor action leading to the collapse of a relatively large part of the structure" (Agarwal and England, 2008).

From the above definition, it can be seen that the term *disproportionate collapse* refers to the extent of the failure area, which, in other words, is the small failure area propagated to a large or uncontrollable area. Progressive collapse is normally referred to as the process of the collapse, which means the structural elements are failing one by one progressively. This means a progressive collapse can occur in a relatively small area without triggering the whole collapse of the building. Therefore, the term *disproportionate collapse* is used more frequently in design guidance, as the major purpose of a design is to avoid the collapse of the building in a large or uncontrollable area.

However, from the author's understanding, in design practice, disproportionate collapse often occurs progressively, and most of progressive collapse will finally cause disproportionate collapse. So, there is no need to differentiate them; thus, in this book, they will refer to the same situation.

1.3 Definition of Robustness

Robustness is another important term in progressive collapse resistance design. Eurocode BS EN 1990 (BSI, 2010, p. 26) provides the definition of *robustness*: "A structure shall be designed and executed in such a way that it will not be damaged by events such as explosion, impact, and the consequences of human errors, to an extent disproportionate to the original cause." Therefore, in the process of structural design, securing the robustness of the structure is an important design task; this is usually overlooked by some of the design engineers. In this book, the detailed method to achieve robustness will be introduced.

1.4 Causes of Progressive Collapse and Collapse Incidents with Different Types of Structures

A progressive collapse can occur as the result of different collapse mechanisms, depending on the load path and structural system, as well as the type, location, and magnitude of the triggering abnormal event. There are different types of triggering events, such as vehicular collision, aircraft impact, and fire and gas explosions. They are examples of the potential hazards and abnormal loads that can produce such an event.

There are several famous examples of progressive collapse incidents due to various triggering events. For building collapse, there is the collapse of the Twin Towers on September 11, 2001, due to aircraft impact. The collapse of World Trade Center 7 later that same day was due to a fire set by the debris of the Twin Towers. The partial collapse of the Ronan Point building was triggered by an internal gas explosion in London in 1968, and a blast induced the partial collapse of the Alfred P. Murrah Federal Building in Oklahoma City in 1995. For space structures, there is the famous collapse incident at the Paris airport. The space frame of the Hartford Civic Center in the United States collapsed in 1978 due to heavy snow. Bridge collapse is another quite common incident; the triggering event can be impact loading from the collision of a ship or overloaded lorries. A recent bridge collapse example is the progressive collapse of the suspension bridge Kutai Kartanegara in East Borneo, Indonesia.

To help readers to fully understand the failure mechanisms of these collapse accidents, these incidents will be explored in detail in Chapters 2 through 6, which will cover the structural system of the collapsed structures, the main reason for the collapse, and possible mitigating methods for preventing similar collapses from occurring in the future.

The above collapse incidents caused the loss of life and financial loss. Therefore, it is our responsibility, as engineers, to tackle this in our design, to deliver a better design to prevent progressive collapse.

1.5 Current Design Guidance for Preventing Disproportionate Collapse

In current design practice, there are several design codes and guidances used worldwide for preventing disproportionate collapse. However, they are mainly for building designs; few are found for bridge and space structure designs. A brief introduction of these guidances is given here.

1.5.1 British and European Design Guidance

The United Kingdom was the first country in the world to publish a design guidance for preventing the disproportionate collapse of buildings. The UK Building Regulations (HM Government, 2013) have led with requirements for avoiding disproportionate collapse. These requirements are refined in material-specific design codes, BS 5950 (BSI, 2001), for structural steelwork. They can be described as following three methods:

1. Prescriptive "tying force" provisions that are deemed sufficient for the avoidance of disproportionate collapse
2. Notional member removal provisions that need only be considered if the tying force requirements could not be satisfied
3. Key element provisions applied to members whose notional removal causes damage exceeding the prescribed limits

According to this design guidance, during the design process, the engineers should make sure that all the structures to be designed comply with BS 5950-1: 2000, Clause 2.4.5, "Structural Integrity" (BSI, 2001). Adequate ties will be incorporated into the frame to reduce the possibility of progressive collapse, as required by the building regulations. Key elements will be designed for sustaining an accidental design loading of 34 kN/m². Eurocode also has the detailed requirement such as Eurocode BS EN 1990 (BSI, 2010), ENV 1991-1-7 (BSI, 2006), and ENV 1991-2-7 (2006). The requirements are similar to those in British design guidance (ENV 1991-1-7 and ENV 1991-2-7).

1.5.2 U.S. Design Guidance

The United States is among the first several countries in the world to publish detailed design guidance for preventing progressive collapse in building design.

Design guidelines for progressive collapse resistant design can be found in several U.S. government documents. The Department of Defense (DOD, 2009) and the General Services Administration (GSA, 2003) provide detailed instructions on design methods to resist the progressive collapse of building structures. Both documents employ the so-called alternate path method (APM) to ensure that structural systems have adequate resistance to progressive collapse. The APM is a threat-independent method. It defines column removal scenarios, which are to forcibly remove the building's columns and analyse the response. It also prescribes the loads for which the damaged structure should be analysed. The demand–capacity ratio (DCR) of each primary and secondary member is calculated to determine the potential for progressive collapse. More details will be given in Chapter 2.

The DOD (2009) methodology is based on the desired level of protection: very low, low, medium, or high. Most building structures fall in the first two categories, and only structures that are mission critical or have unusually high risk fall in the last two categories. Except APM, it also uses the tie force method.

SEI/ASCE 7-05 (ASCE, 2005) is the only general standard in the United States to have a design requirement for progressive collapse. It gives two design methods to resist progressive collapse: direct design method and indirect design method.

The direct design method requires that the resistance to progressive collapse be considered directly during the design process through (1) APM, which seeks to provide an alternate load path after a local failure has occurred, so that the local damage is arrested and major collapse is prevented, and (2) the specific local resistance method, which seeks to provide sufficient strength to resist failure at critical locations.

The indirect design method requires the provision of minimum level of strength continuity and ductility of the structural members. Therefore, an engineer needs to provide structural integrity and design ductile connections to enable the ability of the structure to undergo large deformation and absorb large amount of energy under abnormal loading conditions.

NIST (2007) also gives detailed instructions for reducing the potential for progressive collapse in buildings. It includes an acceptable risk approach to progressive collapse, which involves defining the threat, event control, and structural design to resist postulated events. It also has detailed explanation of the design method, such as the direct and indirect methods and the specific local resistance method (similar to the key element method).

1.5.3 Canadian Design Guidance

CSA-S850 is the only Canadian standard that contains explicit and detailed disproportionate collapse mitigation criteria for buildings (Driver, 2014). It is a standard for the design and assessment of buildings subjected to blast loads. However, it contains provisions for preventing progressive collapse and brittle failure.

The standard limits damage under loads caused by an explosion. It has provisions that aim to prevent the occurrence of post-blast disproportionate collapse, rather than a general disproportionate collapse design guidance.

As a result, these provisions are not based on the so-called threat-independent method introduced in U.S. guidance. The procedures for threat and risk assessment are discussed. Therefore, the potential blast-damaged structure is assessed to determine how the building is expected to be compromised, and this forms the basis for the initiation of the required disproportionate collapse analyses.

1.5.4 Chinese Design Guidance

The Architectural Society of China organized a special committee to compile a design specification for the collapse prevention of buildings (Li et al., 2014). Design and analysis methods for the prevention of earthquake-induced collapse, progressive collapse, fire-induced collapse, and construction error–induced collapse are described in this guidance. The progressive collapse resistance demand is determined based on the energy method, and the earthquake-induced collapse resistance evaluation is based on incremental dynamic analysis.

For fire-induced collapse, the specification requires the structure to resist fire for a sufficiently long time without collapse. Three methods are introduced: the simplified component method, the alternative load path method, and advanced analysis for the entire fire process. The prevention of explosion-induced collapse is mainly achieved by improving the maintenance structures, and there is no specialized explosion prevention design for the main structure.

References

ASCE (American Society of Civil Engineers). 2005. Minimum design loads for buildings and other structures. SEI/ASCE 7-05. Washington, DC: American Society of Civil Engineers.

Agarwal, J., and England, J. 2008. Recent developments in robustness and relation with risk. *Structures and Buildings*, 161(SB4), 183–188.

BSI (British Standards Institution). 2001. Structural use of steelwork in buildings. Part 1: Code of practice for design—rolled and welded sections. BS 5950. London: BSI.

BSI (British Standards Institution). 2010. Eurocode—Basis of structural design: Incorporating corrigenda December 2008 and April 2010. BS EN 1990: 2002 + A1: 2005. London: BSI.

CSA (Canadian Standards Association). 2012. Design and assessment of buildings subjected to blast loads. CSA-S850-12. Ontario: CSA. 810-1.

Driver, R.G. 2014. Canadian disproportionate collapse design provisions and recent research developments. Presented at Structures Congress 2014, Boston, April 3–5.

DOD (Department of Defense). 2009. Design of buildings to resist progressive collapse. UFC 4-023-03. Arlington, VA: Department of Defense, July 14.

GSA (General Services Administration). 2003. Progressive collapse analysis and design guidelines for new federal office buildings and major modernization projects. Washington, DC: GSA.

HM (Her Majesty's) Government. 2013. The Building Regulations 2010: Structure, A3: Disproportionate collapse. Approved Document A, 2004 edition, incorporating 2004, 2010, and 2013 amendments. London: HM Government.

Li, Y., Ren, P.Q., and Lu, X.Z. 2014. Development of the design specification for the collapse prevention of buildings in China. Presented at Structures Congress 2014, Boston, April 3–5.

Ministry of Housing and Local Government. 1968. Report of the inquiry into the collapse of flats at Ronan Point, Canning Town. London: Her Majesty's Stationery Office.

NIST (National Institute of Standards and Technology). 2007. Best practices for reducing the potential for progressive collapse in buildings. Gaithersburg, MD: NIST, Technology Administration, U.S. Department of Commerce.

NIST (National Institute of Standards and Technology) NCSTAR (National Construction Safety Team). 2005. Federal building and fire safety investigation of the World Trade Center disaster. Final report of the National Construction Safety Team on the Collapses of the World Trade Center towers. Gaithersburg, MD: NIST, December.

CHAPTER 2

Progressive Collapse Design and Analysis of Multistorey Buildings

2.1 Introduction

Nowadays, increasingly, the design of multistorey buildings, especially tall building projects, requires more attention to prevent progressive collapse. In the UK, a statuary requirement has been made for the design of buildings above a certain height, and a separate progressive collapse analysis report is required to be submitted and checked by the building control department. This is because tall buildings are more vulnerable in terrorist attacks. As introduced in Chapter 1, the design guidances in the UK and United States, as well as the Eurocode, all give detailed requirements for protecting a building against progressive collapse when designing it. Therefore, in this chapter, the detailed progressive analysis and design method for multistorey buildings is introduced.

For an engineer, it is easier to perform progressive collapse analysis using finite-element packages such as Abaqus® and SAP2000 rather than hand calculation. Therefore, it is imperative for an engineer to have knowledge of the modelling method of progressive collapse analysis. Therefore, in this chapter, a detailed modelling and analysis method of buildings using Abaqus® is demonstrated.

2.2 Progressive Collapse Incidents of Buildings around the World

In this section, several progressive collapse incidents around the world are introduced; what triggered their collapse and the failure mechanisms of each collapse are also discussed.

2.2.1 Ronan Point Collapse

Ronan Point (Ministry of Housing and Local Government, 1968) was a 22-storey tower block in Newham, East London. In 1968, a gas explosion demolished a load-bearing wall, which then triggered the collapse of an entire corner of the building (Figure 2.1).

One of the major reasons a progressive collapse occurs is due to the building's structural system. The Ronan Point tower was built using an old technique known as large panel system (LPS). In this system, all the walls, floors, and stairways were precast. The large concrete prefabricated sections were cast off site and bolted together to construct the building. No further ties were designed for the structure. During the design process, with the exception of the gravity load, the only load action considered was the wind load. Accidental loads such as blast were not taken into consideration (Pearson and Delatte, 2005).

The investigation showed that the explosion was not large, with a pressure of <68.9 kN/m^2 (Levy and Salvadori, 1992). However, due

FIGURE 2.1 Ronan Point collapse. (From http://www.geograph.org.uk/reuse. php?id=2540477. Image copyright © Derek Voller. This work is licensed under the Creative Commons Attribution—Share Alike 2.0 Generic Licence. To view a copy of this licence, visit http://creativecommons.org/licenses/by-sa/2.0/ or send a letter to Creative Commons, 171 Second St., Suite 300, San Francisco, CA 94105.)

to the building's lack of structural redundancy, no alternative load path for the upper floors was designed. The building was designed using building codes with no consideration of potential progressive collapse at that time. Another reason was the poor workmanship of the steel tie plates connecting the walls; the workers failed to tighten the nuts of the connecting studs.

2.2.2 *World Trade Center Collapse*

On September 11, 2001, the Twin Towers (referred to as World Trade Center 1 [WTC1] and World Trade Center 2 [WTC2]) collapsed due to two hijacked aircraft crashing into them (NIST NCSTAR, 2005). This is one of the most famous progressive collapse incidents. Later that day, another building, World Trade Center 7 (WTC7), collapsed at 5:21 p.m. due to the fires set by the falling debris from the Twin Towers (Figure 2.2). WTC7 will be further discussed in Chapter 5.

FIGURE 2.2 World Trade Center collapse. (From http://www.shutterstock.com/pic-83242105.html. Licence granted under ID 83242105 in shutterstock.com. Purchased by CRC Press.)

The towers of WTC1 and WTC2 were designed as so-called tube-in-tube structures, as shown in Figure 2.3. This is one of the major lateral stability systems for tall buildings; it uses closely spaced perimeter columns, along with cores in the centre. Above the 10th floor, there were 59 perimeter columns along each face of the building, and there were 47 heavier columns in the core. There was a large column-free space between the core and perimeter that was bridged by prefabricated floor trusses, as shown in Figure 2.4. The towers also used a conventional outrigger truss between the 107th and 110th floors to further strengthen the cores.

As also shown in Figure 2.4, the floor system consists of 10 cm lightweight concrete slabs with a steel deck supported by the bridging

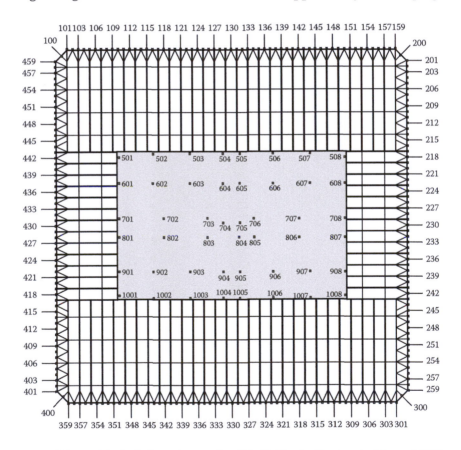

FIGURE 2.3 Typical floor layout for World Trade Center 1. (From NIST NCSTAR, Federal building and fire safety investigation of the World Trade Center disaster, final report of the National Construction Safety Team on the Collapses of the World Trade Center towers, Gaithersburg, MD: NIST, December 2005, Figure 1-3. With permission.)

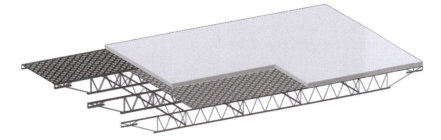

FIGURE 2.4 Typical floor system of the World Trade Center. (From NIST NCSTAR, Federal building and fire safety investigation of the World Trade Center disaster, final report of the National Construction Safety Team on the Collapses of the World Trade Center towers, Gaithersburg, MD: NIST, December 2005, Figure 1-6. With permission.)

trusses and main trusses. This was a typical composite floor system. The trusses were supported on the perimeter by alternate columns. The floors were connected to the perimeter spandrel plates with viscoelastic dampers.

The whole steel frame system, including the steel core and perimeter columns, was protected with sprayed-on fire-resistant material.

A detailed investigation was conducted by the National Institute of Standards and Technology (NIST NCSTAR, 2005). It noted the role of the fires and found that the sagging floors pulled inward on the perimeter columns: "This led to the inward bowing of the perimeter columns and failure of the south face of WTC 1 and the east face of WTC 2, initiating the collapse of each of the towers" (NIST NCSTAR, 2005).

2.3 Minimizing a Threat or Hazard for Potential Collapse

From the above incidents, it can be seen that it is ideal to mitigate the threat of a terrorist attack on buildings from the beginning. However, it is difficult to predict the time and method a terrorist attack will happen.

2.3.1 Threat and Vulnerability Assessment

According to FEMA 426 (DHS, 2003), an engineer can perform a threat or hazard assessment where the threat or hazard of a certain type of building can be identified. After the expected threat or hazard is known, a vulnerability assessment of the building can be performed during the design process. The vulnerability depends on the category of the building, the structural system of the building, the material strength, and so forth.

2.3.2 Risk Assessment

Although threat assessment is possible, there are various triggers of a progressive collapse, such as a vehicle bomb, fire, gas explosion, deliberate attack, vehicle impact, and natural causes; however, due to the limited database of progressive collapse events, it is difficult to assess the probability of occurrence of hazards. Therefore, it is more reasonable to make a risk assessment. Therefore, designers can develop appropriate strategies to mitigate disproportionate collapse in buildings.

As it is stated in IStructE (2013), the main objective of risk assessment is to

1. Identify the hazards to the structure
2. Eliminate the same kind of hazards if it is practical
3. Produce risk reduction measures in the design for the remaining hazards

Risk is the potential for a loss or damage to a structure. The concept of risk involves three components: hazard, consequences, and context (Elms, 1992). The hazard is the triggering event mentioned earlier, such as gas blast or fire. The consequences are the results caused by the hazard, such as a building collapse, personal injury, and loss of life, which must be measured in terms of a value system. Finally, in the decision context, attitudes toward risk and reference made by the engineers will also impact the decisions.

The IStructE guidance (2013) gives the following equation to assess the risk:

$$\text{Risk} = \text{Likelihood} \times \text{Consequence} \qquad (2.1)$$

FEMA 426 (DHS, 2003) explains the risk assessment as threat assessment, asset value, and vulnerabilities (this is the same as the consequences). It also gives the following equation to assess the risk:

$$\text{Risk} = \text{Asset Value} \times \text{Threat Rating} \times \text{Vulnerability Rating} \qquad (2.2)$$

From comparison of the above two equations, it can be seen that the only difference is that Equation 2.1 does not include the asset value; however, in the IStructE guidance (2013), the risk analysis method is based on Table 2.1, where the building has been classified according to the assets value as well. Therefore, these two methods are similar.

Table 2.1 Building Risk Classification

Consequence Class	Building Type and Occupancy
1	Houses not exceeding 4 storeys Agricultural buildings Buildings into which people rarely go, provided no part of the building is closer to another building, or area where people do go, than a distance of 1.5 times the building height
2a Lower risk group	5-storey single-occupancy houses Hotels not exceeding 4 storeys Flats, apartments, and other residential buildings not exceeding 4 storeys Offices not exceeding 4 storeys Industrial buildings not exceeding 3 storeys Retail premises not exceeding 3 storeys of <2000 m^2 floor area in each storey Single-storey educational buildings All buildings not exceeding 2 storeys to which members of the public are admitted and which contain floor areas not exceeding 2000 m^2 at each storey
2b Upper risk group	Hotels, blocks of flats, apartments, and other residential buildings greater than 4 storeys but not exceeding 15 storeys Educational buildings greater than 1 storey but not exceeding 15 storeys Retail premises greater than 3 storeys but not exceeding 15 storeys Hospitals not exceeding 3 storeys Offices greater than 4 storeys but not exceeding 15 storeys All buildings to which members of the public are admitted that contain floor areas exceeding 2000 m^2 but less than 5000 m^2 at each storey Car parking not exceeding 6 storeys
3	All buildings defined above as consequence classes 2a and 2b that exceed the limits on area, number of storeys, or both Grandstands accommodating more than 5000 spectators Buildings containing hazardous substances, processes, or both

Source: BSI, Eurocode 1: Action on structures, Part 1-7: General actions: Accidental actions, BS EN 1991-1-7: 2006, London: BSI, September 2006, Table A.1, p. 34. Permission to reproduce extracts from British Standards is granted by BSI. British Standards can be obtained in PDF or hard-copy formats from the BSI online shop (www.bsigroup.com/Shop) or by contacting BSI customer service for hard copies only (telephone: +44 (0)20 8996 9001, email: cservices@bsigroup.com).

Table 2.2 Occupancy Category (for Building Only)

Nature of Occupancy	Occupancy Category
• Buildings in Occupancy Category I in Table 2-2 of UFC 3-301-01	I
• Low-occupancy buildings[a]	
• Buildings in Occupancy Category II in Table 2-2 of UFC 3-301-01	II
• Inhabited buildings with less than 50 personnel, primary gathering buildings, billeting, and high-occupancy family housing[a,b]	
• Buildings in Occupancy Category III in Table 2-2 of UFC 3-301-01	III
• Buildings in Occupancy Category IV in Table 2-2 of UFC 3-301-01	IV
• Buildings in Occupancy Category V in Table 2-2 of UFC 3-301-01	

Source: DOD, Design of buildings to resist progressive collapse, UFC 4-023-03, Arlington, VA: Department of Defense, July 14, 2009, Table 2-1. With permission from Whole Building Design Guide® (WBDG), a program of the National Institute of Building Sciences.

[a] As defined by UFC 4-010-01, "Minimum Antiterrorism Standards for Buildings."

[b] Occupancy Category II is the minimum occupancy category for these buildings, as their population of function may require designation as Occupancy Category III, IV, or V.

A similar approach is used in DOD (2009), where the risk assessment is carried out based on the category of the buildings, which is determined by the occupancy categories. It comes down to a consideration of consequences. In general, consequences are measured in terms of human casualties, and therefore the occupancy of a building or structure is often the most critical issue. Table 2.2, a reproduction of the table in DOD (2009), gives detailed information of the occupancy categories.

2.4 Design Method

The main objective of a progressive collapse analysis is to determine the potential of structures to collapse and design correspondent mitigation measures to prevent such an occurrence. As introduced in Chapter 1, the major design guidances to be referred to during the design process are GSA (2003), DOD (2009), NIST (2007), FEMA 426 (DHS, 2003), and BS 5950 (BSI, 2001). Both GSA (2003) and DOD (2009) take a so-called threat-independent approach, which means the design target is not to prevent collapse for a specific threat, but to stop the spread of damage after localized damage or collapse has occurred.

In summary, the design guidances in the UK and the United States provide three basic design approaches—one indirect method and two direct methods. These approaches are introduced here.

2.4.1 Indirect Design Method

As mentioned in Chapter 1, the indirect design method requires resistance to progressive collapse through the provision of minimum levels of strength, continuity, and ductility. The indirect method consists of prescriptive design guidance to improve the robustness of the building, including

- Stipulating minimum tying forces for connections
- Identifying key elements
- Designing key elements for increased design loading

This method has been adopted by both BS 5990 (BSI, 2001) and ASCE (2005).

2.4.2 Direct Design Method

The direct design method requires that the resistance to progressive collapse be considered directly during the design process through the alternate path method (APM) and specific local resistance method, which seeks to provide sufficient strength to resist failure at critical locations.

There are two direct design methods. The first one assesses the capacity of the individual members and also the global structure to withstand the loading applied to the structure under each design level hazard. Therefore, it checks their ability to withstand the hazard within agreed acceptance criteria.

The well-known alternate load path is another method which consists of a systematic assessment of the hazards that can be applied to each structural element and a judgement of whether it can withstand that hazard level. If it is deemed that the structural element would not survive the hazard level, then a structural model of the elements and their surrounding structure is developed. This structure is then assessed with the sudden removal of critical structural members, such as columns, to study the structure's ability to redistribute the forces on it without progressively collapsing.

This method has been widely used in the current analysis; most research, such as Fu (2009) and Kim et al. (2009), uses this method. The best way to implement this method is through a finitie element package such as Abaqus® or SAP2000; the detailed modelling method will be introduced in Section 2.7 and Section 2.10.

The General Services Administration (GSA, 2003) provides detailed guidelines on the alternate load path method that enable structural engineers to analyse the ability of a structure to withstand progressive collapse. Readers can further refer to it.

2.4.3 Selection of Design Method

In design guidance such as the British Building Regulations (HM Government, 2013) and DOD (2009), the selection of the above design methods is mainly based on the occupancy category of the building. Table 2-2 in DOD (2009) (as shown in Table 2.3) provides detailed requirements of the progressive collapse analysis method based on the occupancy category, which engineers can refer to. Similar requirements are also provided in the British Building Regulations (2000) and will be introduced in Section 2.4.4.1.

Table 2.3 Occupancy Category and Design Requirements

Occupancy Category	Design Requirements
I	No specific requirements
II	Option 1: Tie forces for the entire structure and enhanced local resistance for the corner and penultimate columns or walls at the first storey
	OR
	Option 2: Alternate path for specified column and wall removal locations
III	Alternate path for specified column and wall removal locations Enhanced local resistance for all perimeter first-storey columns or walls
IV	Tie Forces; alternate path for specified column and wall removal locations; enhanced local resistance for all perimeter first- and second-storey columns or walls

Source: DOD, Design of buildings to resist progressive collapse, UFC 4-023-03, Arlington, VA: Department of Defense, July 14, 2009, Table 2-2. With permission from Whole Building Design Guide® (WBDG), a program of the National Institute of Building Sciences.

2.4.4 Requirements for Robustness of Buildings in Design Guidance

In Sections 2.4.1 and 2.4.2, the major design methods are summarized; they can be classified into indirect design and direct design methods. To help readers fully understand the above methods, in this section, the detailed requirements for robustness in other codes (such as the Eurocode or British code) are explained.

2.4.4.1 British Code In Britain, the Building Regulations (HM Government, 2013) and BS 5950 (BSI, 2001) are the two codes that have detailed requirements for robustness.

2.4.4.1.1 Building Regulations (HM Government, 2013) In Britain, Section 5 of Approved Document A, British Building Regulations (HM Government, 2013), states the clear design methods to reduce the likelihood of a building's disproportionate collapse in the event of an accident.

The key for this guidance is that it requires different levels of robustness requirements based on different consequence classes of building. These classes are determined by the types and sizes of buildings. There are four classes of buildings: Class 1, Class 2a, Class 2b, and Class 3. The building classification method is similar to BS EN1991-1-7 (BSI, 2006), shown in Table 2.1, with only a small difference.

The robustness requirements specified in the Building Regulations (HM Government, 2013) for each class of building are summarized in Table 2.4.

It is required by the Building Regulations (HM Government, 2013) that the key element mentioned in Table 2.4 should be capable of sustaining an accidental design loading of 34 kN/m^2 applied in the horizontal and vertical directions simultaneously (Figure 2.5).

2.4.4.1.2 BS 5950 (BSI, 2001) In BS 5950-1 (BSI, 2001), it is a requirement that all steel structures, irrespective of height or span, comply with prescribed minimum acceptance levels of robustness.

Clause 2.4.2.3 requires that all structures be able to resist notional disturbing forces dictated by the design's gravity load.

Clause 2.4.5 requires that the members be tied together adequately in all directions. Structures that fall outside the restriction on the number of storeys or span (in the case of public buildings) should be appraised for their behaviour under collapse conditions.

Table 2.4 Summary of the Robustness Requirements Specified for Each Class of Building

Consequence Class	Robustness Requirement
1	No additional measures needed.
2a	1. Horizontal ties.
	2. Effective anchorage of suspended floors to walls in accordance to code of practice.
2b	1. Horizontal ties or effective anchorage of suspended floors to walls in accordance to code of practice.
	2. Check that with the removal of a column or wall, the building remains stable and the area at any storey at risk of collapse does not exceed 15% of the floor area of that storey or 100 m², whichever is smaller (see Figure 2.5). If the collapse area limit is exceeded, such an element should be designed as a key element.
3	Systematic risk assessment of the building should be taken into account for all critical situations, and correspondent design measures should be taken.

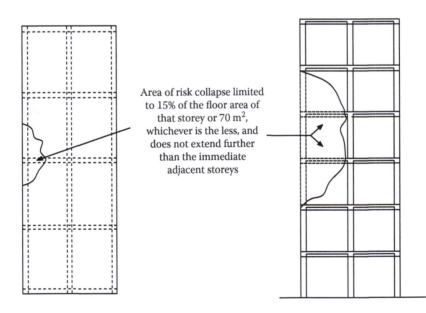

Area of risk collapse limited to 15% of the floor area of that storey or 70 m², whichever is the less, and does not extend further than the immediate adjacent storeys

FIGURE 2.5 Area at risk of collapse in the event of an accident. (From HM Government, The Building Regulations 2010: Structure, A3: Disproportionate collapse, Approved Document A, 2004 edition, incorporating 2004, 2010, and 2013 amendments, London: HM Government, 2013, Diagram 24. Permission to reproduce from HM Government. Copyright © Crown Copyright, 2013.)

Clauses 2.4.5.1 and 2.4.5.2 specify robustness-specific requirements that must also be satisfied by all structures.

BS 5950-1 (BSI, 2001) also requires that a building be capable of sustaining an accidental design loading of 34 kN/m². Other requirements have been introduced in Chapter 1 and therefore are not repeated here.

2.4.4.2 Eurocode

2.4.4.2.1 Eurocode BS EN 1990 (BSI, 2010) In Eurocode BS EN 1990 (BSI, 2010), there are several basic requirements for the design of structures, including that the structure shall "sustain all actions and influences likely to occur during execution and use" and "be designed to have adequate structural resistance, serviceability and durability."

As introduced in Chapter 1, Eurocode BS EN 1990 (BSI, 2010) has particular relevance to structural robustness.

2.4.4.2.2 ENV 1991-1-7 (BSI, 2006) Eurocode BS 1991-1-7 (BSI, 2006) proposes three approaches to design for accidental actions, such as impact and explosions, with each assigned to a different category of accidental design situations.

Category 1 is defined as having limited consequences. No specific considerations for accidental loads are required.

Category 2 has medium consequences and requires either a simplified analysis by static equivalent models or the application of prescriptive design or detailing rules, depending on the specific circumstance of the structure in question.

Category 3 relates to large consequences, recommending a more extensive study, using dynamic analyses, nonlinear models, and load structure interactions, if appropriate.

Eurocode BS 1991-2-7 (2006) also has the below requirements in Clause 3.3, "Accidental Design Situations—Strategies for Limiting the Extent of Localised Failure":

(1) In the design, the potential failure of the structure arising from an unspecified cause shall be mitigated.

(2) The mitigation should be reached by adopting one or more of the following approaches:

- Designing key elements, on which the stability of the structure depends, to sustain the effects of a model of accidental action *Ad*;

- Designing the structure so that in the event of a localised failure (e.g. failure of a single member) the stability of the whole structure or of a significant part of it would not be endangered;
- Applying prescriptive design/detailing rules that provide acceptable robustness for the structure (e.g. three-dimensional tying for additional integrity, or a minimum level of ductility of structural members subject to impact).

2.5 Structural Analysis Procedures

In this section, the structural analysis for disproportionate collapse will be introduced.

2.5.1 *Four Basic Analysis Procedures*

Four main analysis procedures based on column removal are included in GSA (2003):

- Linear static
- Linear dynamic
- Nonlinear static
- Nonlinear dynamic

The procedures are based on the category of the buildings, with consideration of their degree of structural regularity. The implementation of these four procedures are mainly through analysis software such as Abaqus® SAP 2000, which will be introduced in Section 2.7.

2.5.2 *Load Combinations*

Upon performing analysis, it is recognized that under certain column removal scenarios, the loading present will be less than the ultimate design loading for the structure; therefore, GSA (2003) recommends that the load present on the column removal should be the full dead load (DL) plus 0.25 times the live load (LL). However, the structural performance under a possible column removal is influenced by the dynamic characteristics of the structure itself, such as the dynamic amplification effect, which is automatically captured for dynamic procedures but not for static analyses. Therefore, it is recommended that for static procedures, the load applied to the structure is increased to two times the value recommended above.

The required load combinations for different analysis procedures are summarized in Table 2.5.

Table 2.5 Load Combination for Different Analysis Procedures

	Load Combination
Static linear	2 × (Dead + 0.25 Live)
Static nonlinear	2 × (Dead + 0.25 Live)
Dynamic linear	Dead + 0.25 Live
Dynamic nonlinear	Dead + 0.25 Live

2.5.3 Energy Method in Progressive Collapse Analysis

In recent years, the energy method (Szyniszewski, 2009) has been used by some researchers for progressive collapse analysis. Therefore, it is worth introducing here.

It is a physics-based collapse simulation with an emphasis on the development of energy flow relationships. It has been proposed that energy flow during progressive collapse can be used in the evaluation of building behaviour and localized failure. If a collapsing structure is capable of attaining a stable energy state through absorption of gravitational energy, then collapse will be arrested. Otherwise, if a deficit in energy dissipation develops, the unabsorbed portion of released gravitational energy is converted into kinetic energy and collapse propagates from unstable state to unstable state until total failure occurs. Therefore, the energy absorption can be used to make a judgement on structural behaviour in structural members.

2.6 Acceptance Criteria

In DOD (2009), the acceptance criteria to be used are dependent on the type of analysis adopted. The acceptance criteria for the four analyses are summarized in Table 2.6.

2.6.1 Acceptance Criteria for Linear Procedures

In DOD (2009), for linear elastic procedures, the acceptance criteria are based on the demand–capacity ratio (DCR) for each structural member or connection in the design. The use of the linear elastic procedures is limited to structures that meet the requirements for

Table 2.6 Acceptance Criteria for Different Analysis Procedures

	Acceptance Criteria
Static linear	Demand–capacity ratio
Static nonlinear	Rotation/ductility
Dynamic linear	Demand–capacity ratio
Dynamic nonlinear	Rotation/ductility

irregularities and DCRs. The DCR is simply the force in each member or connection under the considered scenario divided by the expected ultimate, unfactored capacity of the member or connection considered. It is calculated as follows:

$$DCR = QUD/QCE \qquad (2.3)$$

where,

QUD is the resulting action (demand) determined in the component or connection or joint (moment, axial force, shear, and possible combined forces), and

QCE is the expected strength of the component or element, as specified in Chapters 4 through 8 in DOD.

Using the DCR criteria of the linear elastic approach, structural elements and connections that have DCR values that exceed the following allowable values are considered to be severely damaged or collapsed. The allowable DCR values for primary and secondary structural elements are

- DCR < 2.0 for typical structural configurations (GSA, 2003, Section 4.1.2.3.1)
- DCR < 1.5 for atypical structural configurations (GSA, 2003, Section 4.1.2.3.2)

Figure 4.3 of GSA (2003) contains the DCR limit calculation table for beams and columns. Readers can refer to it to choose the correct DCR in the design.

2.6.2 Acceptance Criteria for Nonlinear Procedures

As shown in Table 2.6, for nonlinear procedures, the acceptance criteria are based on less onerous rotation and ductility demands for the members and connections considered. DOD (2009) gives the detailed rotation and ductility requirements based on the different structural materials (such as steel or concrete) and different types of structural members (beams or slabs). Refer to DOD (2009) for detailed guidance.

2.7 Progressive Collapse Analysis Procedures Using Commercial Programs

It can be seen that the recommendations for each type of analysis result in different degrees of conservatism, which depend on the structural form considered. With normal computer software, the linear static procedure is easy to implement; the nonlinear dynamic procedure requires more modelling techniques, and therefore advanced modelling programs such as SAP2000 and Abaqus®. Here the two major procedures, linear static and nonlinear dynamic, will be introduced which are based on DOD (2009) and GSA (2003).

2.7.1 Analysis Procedure for Linear Elastic Static Progressive Collapse Analysis

- Develop a model in an analysis program.
- Remove a vertical support from the location being considered and conduct a linear static analysis of the model of the structure with a load combination of $2(DL + 0.25LL)$ according to Table 2.5.
- Check members and connections with DCR values that exceed the acceptance criteria provided in GSA (2003, Section 2.7).
- If the DCR for any member end or connection is exceeded based on shear force, the member is considered a failed member. If the flexural DCR values for both ends of a member or its connections, as well as the span itself, are exceeded, the member is considered a failed member.
- Remove the failed member from the model. All dead and live loads associated with failed members should be redistributed to other members in adjacent bays.
- For a member or connection with a QUD/QCE ratio exceeding the applicable flexural DCR values, place a hinge at the member end or connection to release the moment. (This can be achieved using an analysis program such as SAP2000.)
- Rerun the analysis and repeat the steps until no DCR values are exceeded.
- Judgement of the collapse of buildings. If moments have been redistributed throughout the entire building and DCR values are still exceeded in areas outside of the allowable collapse region, the structure is considered to have a high potential for progressive collapse.

It can be seen that the linear elastic static analysis procedure is simple and straightforward, but the steps need to be repeated several times to reach the conclusion. Readers can fulfil this analysis through programs such as SAP2000, ETABS, or STAAD Pro.

2.7.2 Analysis for Nonlinear Dynamic Procedure

- Develop a model with a load combination of full dead load plus 0.25 times the live load according to Table 2.5.

- Use a static step to determine the forces under this load combination: full dead load plus 0.25 times the live load.
- Remove a column in a dynamic time step that is less than 1/10 the period of the fundamental mode.
- Determine the subsequent free dynamic response under the column removal scenario, including geometric nonlinearity (use an analysis program such as Abaqus® or ANSYS), or insert the plastic hinges (use a program such as SAP2000).
- Determine the maximum forces, displacements, and rotations for each of the members or connections involved in the scenario from the dynamic analysis for comparison with the acceptance criteria outlined in the GSA guidelines (2003, Table 2.1).

2.8 Collapse Mechanism of Buildings

It is necessary to understand the collapse mechanism in the design. In this section, the collapse mechanism will be discussed. According to Ettouney (2004), the process of a progressive collapse for a multi-storey building consists of the following phases:

- Loss of target column
- Loss of adjacent columns
- Structural instability
- Collapse of the structure

2.8.1 Catenary Action

In the process of collapse, as shown in Figure 2.6, the catenary action helps to carry the vertical load through the axial force of a horizontal member after the deflection of the member has become significant. Due to the large deflection, the member will stretch, which leads to plastic stretching and bending. According to the plastic theories, increasing the axial force decreases the plastic moment capacity of the member, which has to be taken into account. The catenary action can also be used for members without any moment capacity, and therefore also for beams without moment resistant connections.

In designing a building to prevent progressive collapse, catenary action plays an important role in progressive collapse resistance, as it increases the capacity of a frame after the removal of columns. Therefore, it should be taken into consideration in the design.

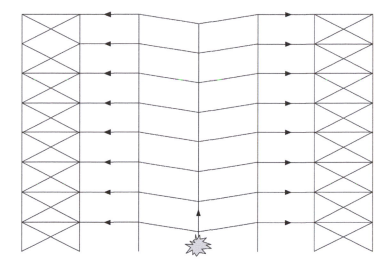

FIGURE 2.6 Catenary action.

2.8.2 Collapse Mechanism of Steel Composite Frames

Guo et al. (2013, 2015) investigated the collapse mechanism of steel composite frames through experimental studies. Both rigid and semirigid composite connections were investigated in their test. They found that when a column is removed, the frame becomes deformed in a six-phase process: elastic, elastic–plastic, arch, plastic, transient, and catenary phases.

In the first phase (elastic), the load–deformation relationship of the frame is linear, as the specimen is almost elastic and the deformation is small. In the second phase (elastic–plastic), the load increases nonlinearly with the increase of displacement; meanwhile, the stiffness of the curve decreases.

In the third phase (arch), the curve presents an arch trend. The resistance of the frame increases until it reaches the peak value, when it starts to reverse. The arch action is advantageous to the performance of a frame subjected to column loss and can provide a higher resistance, so-called peak resistance, which is higher than plastic resistance.

During the fourth phase (plastic), the yielding of the structural members starts to develop and the resistance of the frame begins to rise.

In the fifth phase (transient), the load-bearing mechanism of the frame transforms from a plastic hinge action to catenary action.

After a slight declination in the stiffness of the frame, the curve goes into the final phase (catenary). During this phase, the vertical load is sustained by catenary action. The loss of the moment resistance in the joints ceases the plastic hinge action. The slab reinforcement and steel beam provide the tensile force caused by the catenary action. The vertical load increases linearly with the increase of vertical displacement.

2.8.3 *Collapse Mechanism of Concrete Buildings*

Different to steel composite buildings, reinforced concrete is the major material of the concrete buildings. The research also indicated that the catenary action plays an important role during the collapse process. Sasani and Kropelnicki (2008) conducted an experiment to study the behaviour of beams bridging over the removed column. A 3/8 scale test of the second floor beam, bridging the removed column, was constructed with fixed boundary conditions. The rebar fracture was observed on bottom side of the beam at the location of removed column due to the increase of gravity loading caused by the removed column. It is also observed that, following the bar fractures, catenary action provided by the top reinforcement results in the increasing resistance of the beam. When vertical displacement increases to 0.22 m, the top continuous bars at the center of the beam which were previously in compression yielded in tension, this is resulted from the catenary actions.

For columns, if the buildings are in the seismic zones, in the earthquake design, beam depth and reinforcement detailing were chosen to create a so-called weak-column strong-beam mechanism as well as to reduce joint shear stresses. In addition, most columns are also required to design in a ductile manner. This results concentration of damage in the non-ductile columns, which may causes the axial collapse of these columns when the adjacent column is destroyed or due to the heave earthquake. In earthquake design short column is also prevented to avoid shear failure in the columns. If the building is not in the seismic zone, ductile design is also required by most of the design guidance, therefore, brittle shear failure rarely happens, axial and flexural failure can be expected in the columns.

2.9 Mitigating Measures in Design of Multistorey Buildings against Progressive Collapse

Depending on the importance, structural type, and location of the building, there are various measures to prevent or delay the collapse of the building. These measures are introduced in detail in this section. However, some strategies involve a high cost of construction. Due to constraints from clients, engineers should make correct judgements and find a cost-effective way to design a building to prevent progressive collapse.

2.9.1 Hazard Mitigation

As discussed in Section 2.1, although it is hard to assess and mitigate the hazards, there are also some design measures that can be used to reduce the hazard and minimize the likelihood and magnitude of the threat.

This can be achieved by adopting preventive measures that discourage or impede an attack. Measures such as the following can be used:

- Have a Ballard surround the building. This will minimize the chance for a vehicle collision with explosives.
- Have security checks at entrance. This will minimize the possibility of explosives being brought into the building.
- Providing the occupants with either a safe area or an effective escape route and assembly area.

Readers can refer to FEMA 426 (DHS, 2003) for further guidance.

2.9.2 Alternative Path Method

DOD (2009), GSA (2003), and BS 5950 (BSI, 2001) all recommend the alternative path method (APM) in progressive collapse design. It is a threat-independent methodology, meaning that it does not consider the type of triggering event or the reason for the damaged condition, but concentrates on the response of the structure after the triggering event has destroyed critical structural members. BS 5950 states, "The Alternative Path approach must show that the structure is capable of bridging over a notionally-removed column or a notionally-removed section of wall."

In the design process, if one component fails, alternate paths should be designed that are available for the load so that a general collapse does not occur. The alternative path method transfers the forces through the lost element to other structural elements. This method is generally applied in the context of a missing column scenario to assess the potential for progressive collapse and is used to check whether a building can successfully absorb the loss of a critical member. The technique can be used for the design of new buildings or for checking the capacity of existing structures.

The advantage of this method is that it supports structural systems with ductility, continuity, and energy-consuming properties that are suitable in preventing progressive collapse. This approach would certainly discourage the use of a large transfer girder or suspension floors hanging over the roof truss.

This method is consistent with the seismic design approach that promotes the regular structures that are well tied together. They also require ductile details so that plastic rotations can take place (NIST, 2007).

2.9.3 Protecting Key Elements to Prevent Local Failure

This is the so-called key element method we have discussed several times earlier. The designer needs to firstly identify the key elements, therefore, enhancement can be made in the design for these key elements. Any member upon which significant proportions of a structure rely for stability and support should be designed as a key element. Therefore, the major key elements could be transfer beams, hangars, and trusses that are supporting a large-load tributary area. Therefore, a robust design can be achieved by satisfying a number of requirements regarding the location of principal bracing elements for sway resistance, tensile resistance of column splices, and anchoring of floor units. For example, a non-sway structure with a bracing system will collapse if the bracing system is destroyed by a localized blast or accidental loading.

An accurate way to identify the key element is through the removal of certain structural elements. It is required by BS 5950 (BSI, 2001) that "structures should be inherently capable of limiting the spread of local failure regardless of the cause and ideally should be capable of locally bridging over a missing member—albeit in a substantially deformed condition." In this case, the missing member can be any single column or beam carrying a column. If it is found that removal of a member results in damage more extensive than the specified limits, the member must be designed as a key element.

If the absolute safety and integrity of a structure is to be guaranteed, key elements should be designed to resist all foreseeable, abnormal loading conditions. However, to consider the consequences of all these conditions would be an uneconomic design. As mentioned earlier, the Building Regulations (HM Government, 2013) require that an accidental load up to 34 kN/m^2 is used to check the adequacy of the key element.

2.9.4 Tying Force Method

DOD (2009), NIST (2007), and the British Building Regulations (HM Government, 2013) all suggest the tying force method (TFM). In this method, the building is designed to be tied together, using ties, therefore enhancing the continuity, ductility, and development of alternate load paths. The TFM is an indirect design approach. It requires that a minimum tie force capacity be made available in the system to transfer loads from a damaged part to the remainder of the structure, therefore ensuring adequate resistance to a disproportionate collapse. This method requires the minimum amount of design effort.

For a very low level of protection, it is sufficient to provide a prescribed horizontal tie force capacity.

For a low level of protection, both the horizontal and vertical tie capacities have to be provided. If an adequate vertical tie capacity is not present, then APM is required.

When the objective is to achieve medium or high levels of protection, structures have to be designed for the prescribed horizontal and vertical tie forces, should satisfy the minimum ductility requirement, and should additionally be checked by APM for specific damage scenarios. In all cases, APM is permitted only if the horizontal tic capacity is present.

DOD (2009) requires that three horizontal ties must be provided: longitudinal, transverse, and peripheral. Vertical ties are required in columns and load-bearing walls. Figure 2.7 illustrates these tie requirements for frame construction.

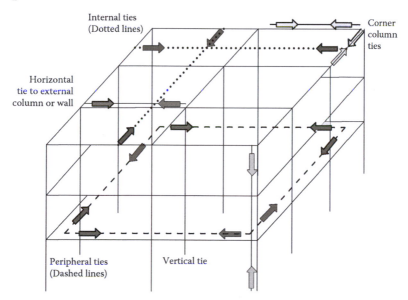

FIGURE 2.7 Tie forces in a frame structure. (From DOD, Design of buildings to resist progressive collapse, UFC 4-023-03, Arlington, VA: Department of Defense, July 14, 2009, Figure 3-1. With permission from Whole Building Design Guide® (WBDG), a program of the National Institute of Building Sciences.)

The TFM relies on the formation of catenary action to mitigate collapse; however, in the design, it is difficult to determine how much catenary action will take place in response to a specified event, or alternatively, how much must be developed to mitigate collapse. Therefore, the accuracy of the criteria and their appropriateness for adoption are questionable at this time.

DOD (2009) provides the below formula to calculate the required tie strength in the design:

$$\Phi R_n \geq \Sigma \gamma_i Q_i \qquad (2.4)$$

where

ΦR_n = design tie strength,

Φ = strength reduction factor,

R_n = nominal tie strength calculated with the appropriate material-specific code, including the overstrength factors from Chapters 5 through 8 of ASCE 41.

$$\Sigma \gamma_i Q_i = \text{Required Tie Strength}$$

where

γ_i = load factor

Q_i = load effect.

2.9.5 *Increasing Structural Redundancy*

When designing a building, structural redundancy is an effective way to ensure the alternate load paths when a structural member fails. The use of redundant lateral and vertical force resisting systems is highly encouraged when considering progressive collapse. However, the redundancy will increase the overall cost; therefore, depending on the category of the building, one should make an engineering judgement on the version of redundant structures.

FEMA 369 (Building Seismic Safety Council, 2000) highlights the need to design more redundant structures so that alternate load paths are available in the event of local failure and the structure retains its integrity and continues to resist a lateral earthquake load. Additional redundancy in framed structures is to be provided by incorporating moment resisting joints in the vertical load carrying system and providing different types of seismic force resisting systems, where a backup system can prevent catastrophic effects if distress occurs in the primary system. The increase in redundancy is considered to be a function of moment resisting frame placement and the total number of such frames. Though this guideline has no specific criteria to design for progressive collapse, the requirement for redundancy can be referred to in progressive collapse resistance design.

2.9.6 *Utilizing the Ductility*
In preventing progressive collapse, it is ideal to design both primary and secondary structural elements capable of deforming well beyond the elastic limit. An engineer needs to design ductile construction materials for both structural members and connections. One should avoid low-ductility detailing in elements that might be subject to dynamic loads or very large distortions during localized failures.

For concrete structures, the ductility is secured by placing continuous bottom reinforcement over supports, providing sufficient confinement at joints, and having adequate ties to allow for load transfer, peripheral ties at the spandrels, internal ties through floor slabs and beams, horizontal ties to columns and walls, vertical ties along the perimeter structure, and tension ties for precast concrete construction.

For steel structures, this can be achieved through a moment resisting frame at the critical location and avoiding splices at critical column locations, or using moment resisting splices instead.

2.9.7 *Connection Strengthening and Detailing*
The ability of structures to develop a tie force or an alternate path to resist disproportionate collapse relies greatly on the structural detailing of the beam-to-column connections. For an example, if a beam is subject to large deformations, the load will be resisted partly by axial or membrane tension, which will also bring a large axial force to the connections. Therefore, particular care should be taken in the beam-to-column connection design to ensure the connections are able to accommodate the extra force.

For concrete structure, GSA (2003) also has the requirement to ensure beam-to-beam continuity across a column. At the connections, the connection reinforcement should also be detailed to behave in a ductile manner.

Guo et al. (2015) found that for a multistorey building, the tension zone of the connection should be reinforced or enhanced to prevent disproportionate collapse. They provided several connection enhancement methods:

Welding reinforcement. In this method, an 8 mm leg-size welding seam is added to the bottom of the endplate. The property of the welding seam is the same as that for the steel beam.

Haunch reinforcement. Welding a triangle haunch beneath the beam has been shown to be very effective for repair, rehabilitation, or new construction. In Appendix D of GSA (2003), a rigid steel connection with a haunch is recommended for use in designs to prevent progressive collapse.

Angle-steel reinforcement. In this method, an angle unit is added beneath the steel beam, whose tips are welded to the flanges of the beam and column, respectively. Under service loads, the angle unit is not involved in load resistance. Following an increase of the joint rotation, angle steel is straightened under tensile stress (the load is the dynamic load caused by the removal) and involved in load resistance gradually. Compared to the haunch and other reinforcing methods, this method does not influence the design under service loads, but it does improve the robustness of the joint to prevent failure.

2.9.8 Preventing Shear Failure

Since shear failure is brittle, GSA (2003) also requires that sufficient strength and ductility should be designed for primary structural members in the progressive collapse design to prevent shear failure.

2.9.9 Use of Steel Cables to Prevent Progressive Collapse

Tan and Astaneh-Asi (2003) proposed a method to prevent progressive collapse by placing a series of steel cables inside the floor reinforced concrete (RC)–steel deck slab and anchoring the end of the cables to braced or shear wall bays for new construction. For existing buildings, the cables are placed along the existing beams and anchored to the columns. It was found from their experimental tests that the cables acting in a catenary action mode added to the strength of the system and prevented progressive collapse of the floor after removal of the middle column. The mechanisms performed well in the tests and proved to be very efficient and economical in preventing progressive collapse of the tested specimen.

Muhammad and Thaer (2012) proposed a scheme that is comprised of placing on each floor vertical cables connected at the ends of beams. The cables are hung on a steel-braced frame seated on top of the building. In the case of a column collapse, the cables transfer the residual loads above the failed column to the roof-braced frame.

The above two methods provide a backup system for resisting the progressive collapse of buildings; however, they also increase the overall cost of the building. Therefore, an engineer should make his or her selection based on the category of the building and the cost requirements of the client and provide more efficient and cost-effective design solutions.

2.10 Case Study: Progressive Collapse Analysis of World Trade Center 1 Using Abaqus®— Nonlinear Dynamic Procedure

In this section, a modelling example of progressive collapse analysis using Abaqus® is demonstrated. The purpose of this modelling example is to demonstrate how to perform the column removal analysis prescribed in both UK and U.S. codes, rather than simulating the collapse process of the World Trade Center; therefore, in the analysis, only two columns at the level where the aircraft collided are removed. The model replicates the original structure of the World Trade Center 1, based on Figure 2.3 and other drawings from NISI (2005). In the analysis, reasonable simplifications are also made.

2.10.1 Prototype Building
The WTC1 was chosen as the prototype building to demonstrate the analysis procedures. The WTC1 is first set up using the three-dimensional (3D) modelling program ETABS, as shown in Figure 2.8. Then,

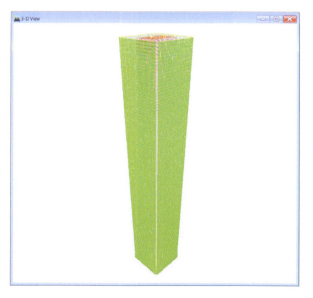

FIGURE 2.8 3D ETABS model of WTC1. (ETABS screenshot reprinted with permission of Computer and Structures.)

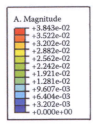

FIGURE 2.9 3D Abaqus® model of WTC1. (Abaqus® screenshot reprinted with permission from Dassault Systèmes.)

using the program designed by the author (Fu, 2009), the model is converted into Abaqus® INP files and is analyzed. The Abaqus® model is shown in Figure 2.9.

2.10.2 Modelling Techniques
All the beams and columns are simulated using *BEAM elements from the Abaqus® library. The structural beam elements

are modelled close to the centreline of the main beam elements. The slabs are simulated using the four-node *SHELL element. Reinforcement was represented in each shell element by defining the area of reinforcement at the appropriate depth of the cross section using the *REBAR element from the Abaqus® library. This reinforcement is defined in both slab directions and was assumed to act as a smeared layer. The beam and shell elements are coupled together using rigid beam constraint equations to give the composite action between the beam elements and the concrete slab. The material properties of all the structural steel components were modelled using an elastic–plastic material model from Abaqus® library, which incorporates the material nonlinearity. The concrete material was modelled using a concrete damage plasticity model from Abaqus®. The tensile strength of the concrete is ignored after concrete cracking. The shell elements are integrated at nine points across the section to ensure that the concrete cracking behaviour is correctly captured. The models are supported at the bottom, as shown in Figure 2.9. The mesh representing the model has been studied and is sufficiently fine in the areas of interest to ensure that the developed forces can be accurately determined. The steel beam to the column connections is assumed to be fully pinned. The continuity across the connection is maintained by the shell element acting across the top of the connection. Therefore, the beam to the column connection is more or less like a semirigid composite connection that simulates the characteristics of the connections in normal construction practice. Detailed modelling techniques are explained in Fu (2009).

2.10.3 Load Combination and Column Removal

The nonlinear dynamic procedure is used here. Therefore, the load combination is $DL + 0.25LL$, as required by the acceptance criteria outlined in GSA (2003, Table 2.1). The APM is applied here to perform progressive collapse checking of WTC1. The resistance ability of the building under sudden column loss is assessed using the nonlinear dynamic analysis method with the 3D finite-element technique available in Abaqus®. The columns to be removed are forcibly removed by instantaneously deleting them. As highlighted in Figure 2.10, in the analysis, two columns at the floor level where the aircraft collided into the WTC1 were removed. The removal analysis is performed and the analysis result is demonstrated here. In the analysis, 0.05 damping ratio is selected.

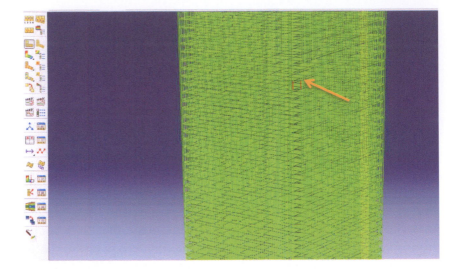

FIGURE 2.10 The two columns removed in the model. (Abaqus® screenshot reprinted with permission from Dassault Systèmes.)

2.10.4 Major Abaqus® INP File Commands Used in the Simulation
Using the program of Fu (2009), the INP file consists of several main parts. Readers can refer to the Abaqus® manual for detailed examples; here only the most important parts are explained in detail.

1. Define the node coordinates (determine the coordinates of the building).

```
*node,nset = Nodes (defining a node set called nodes)
1,2.1,0,3.6
2,60.9,0,3.6
3,63.42,2.284782,3.6
4,63.42,61.08478,3.6
5,61.22213,63.51,3.6
6,2.422131,63.51,3.6
. . . . . . . .
*ncopy,old set = Nodes,new set = DECKlevel, change
number = 100000,shift
0.,0.,0.25
0
```

2. Define the shell elements (define the slabs and walls).

```
*element,type = s4r,elset = WALL1 (The first command
line defines the shell element type and slab name)
```

```
100001,35488,35723,36046,35811 (The second line
defines the shell name and four notes for each
shell.)
. . . . . . .
*element,type = s4r,elset = DECK1
100052,35812,36077,36078,35813
. . . . . . .
```

3. Define the frame element (define the beam, column, and bracing).

```
*element,type = b31,elset = FPERIMETER1
1,35483,35806
*beam section,section = Box,
elset = FPERIMETER1,material = steel
0.19,0.16,0.02,0.02,0.02,0.02
0,1,0
. . . . . .
```

4. Define the boundary condition (choose all the nodes at the bottom and name them bottomnode).

```
*nset,nset = bottomnode
546
545
. . . . . . .
*boundary (defining the boundary condition).
bottomnode,1,6
```

5. Connect the beam to the slab (use the constraint equation to connect the slab to the beam to make a composite action).

```
*nset,nset = allbeam,elset = allbeam
*ncopy,old set = allbeam,new set = sbeam,change
number = 100000,shift
0.,0.,0.25
0.
*mpc (constraint equation is defined here)
beam,sbeam,allbeam
```

6. Release definition (define the moment release for the beams).

```
*elset,elset = momrels1
54228
. . . . . . .
*elset,elset = momrels2
54230
```

```
. . . . . . .
*release
momrels1,s1,M1-M2 (to simulate the pin connection)
momrels2,s2,M1-M2 (to simulate the pin connection)
```

7. Shell section definition (define the shell element for concrete walls and slabs).

```
*shell section,elset = wall1,material = Concrete
0.45,9
*rebar layer (here the reinforcement in the shell
elements are defined)
a252x,50.26e-6,0.200,0.03,s460,,1
a252y,50.26e-6,0.200,0.03,s460,,2
*shell section,elset = DECK1,material = Concrete
0.4,9
*rebar layer
a252x,50.26e-6,0.200,0.03,s460,,1
a252y,50.26e-6,0.200,0.03,s460,,2
```

8. Define the element set for column removal (define an element set named removal; choose two elements with IDs 8839 and 9141).

```
*elset,elset = removal
8839,9141
```

9. Define the materials for the steel and concrete.
   a. Define the steel material (this sets up the material property for the steel member).

   ```
   *MATERIAL,name = Steel
   *elastic
   2.10E+11,0.3
   *plastic
   355.000e6,0.000,20.
   460.800e6,0.182,20.
   *density
   7850
   . . . . . . .
   ```

   b. Define the concrete material (this sets up the material property for the concrete member).

   ```
   *material,name = Concrete
   *elastic
   14331210191,0.2,20
   *Concrete Damaged Plasticity
   ```

```
30,,1.16,,0.
*Concrete Compression Hardening
28662420 ,   0       ,,   20
30000000 ,   0.0005 ,,   20
```

10. Define the analysis steps. Three analysis steps will be defined here. The first is the static step, which applies the normal gravity load, such as the dead and live loads, to the structure. The second is the column removal step, and the third is the dynamic response step, which records the dynamic response after the column has been removed. They are demonstrated as follows:

a. Static step

```
*STEP,INC = 5000,nlgeom = yes,unsymm = yes
*STATIC
0.25
*controls,analysis = discontinuous,field =
  displacement
*controls,parameters = field,field =
  displacement
0.01,1.0
*controls,parameters = field,field = rotation
0.02,1.0
*dload
all,grav,9.81,0,0,-1 (self-weight load)
DECK1,p,-0.5e3 (Live load)
*restart,write,FREQ = 100
*output,field,frequency = 2
element output,elset = frame (defining the output
  request for frame elements, however, the elset
  called frame need to be defined first)
S
E
PE
element output,elset = DECK1 (defining the
  output request for slab elements)
1,3,5,7,9
S
E
PE
PEEQ
PEEQT
*element output,rebar = a192x,elset = DECK1
  (defining the output request for slab rebars)
rbfor
```

```
*element output,rebar = a192y,elset = DECK1
  (defining the output request for nodes)
rbfor
*node output
u
v
a
cf
rf
nt
*end step
```

b. Column removal step (the two columns chosen in part 8 are removed here)

```
*step,inc = 5000
*dynamic,haftol = 100000000,initial =
  no,alpha = -0.05 (choose the default damping
  ratio as -0.05)
0.0025,0.02,0.0000001,0.01
*model change,remove
removal
*end step
```

c. Column removal step (the dynamic response of the structure is captured in this step)

```
*step,inc = 5000
*dynamic,haftol = 1000000000,initial =
  no,alpha = -0.05
0.0025,2.0,0.0000001,0.05
*end step
```

2.10.5 Modelling Result Interpretation

2.10.5.1 Contour Plots   The modelling result can first be checked in terms of the contour plots. In the *Ribbon*, click on *Result*, and then choose *Field Output* and the *Primary Variable* to investigate, such as vertical displacement *U3*. Figure 2.11 shows the contour plot of the vertical displacement.

2.10.5.2 Time History of Certain Parameters   The time history of any parameter one wants to investigate can be plotted following the below procedures:

- Go to *Result* and click on *XY Date*, as shown in Figure 2.12.
- A new window will pop up (Figure 2.13).

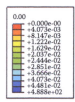

FIGURE 2.11 Contour plot of vertical displacement. (Abaqus® screenshot reprinted with permission from Dassault Systèmes.)

- Click on *ODB Field Output* and select the result parameters you want to investigate, such as *Axial Force SF1* (Figure 2.14).
- Choose the element (Figure 2.15) from the model.

After selection, click on *Plot*. You can get the result of the axial force (we chose section force *SF1*, which represents the axial force), which can also be exported to an Excel file (see Figure 2.17).

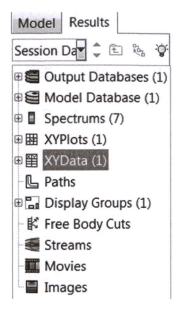

FIGURE 2.12 Selecting the XY date. (Abaqus® screenshot reprinted with permission from Dassault Systèmes.)

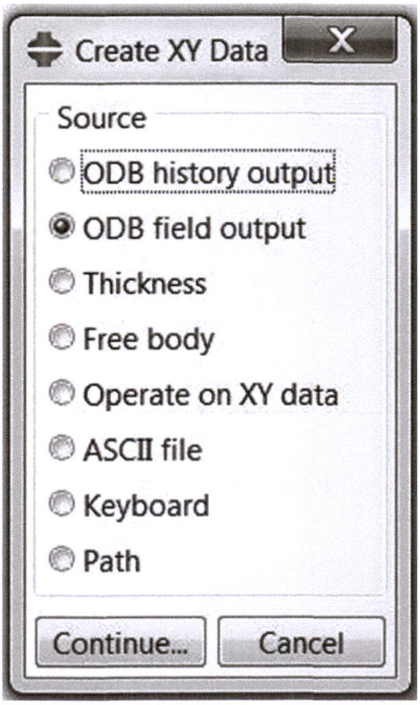

FIGURE 2.13 Selecting ODB field output. (Abaqus® screenshot reprinted with permission from Dassault Systèmes.)

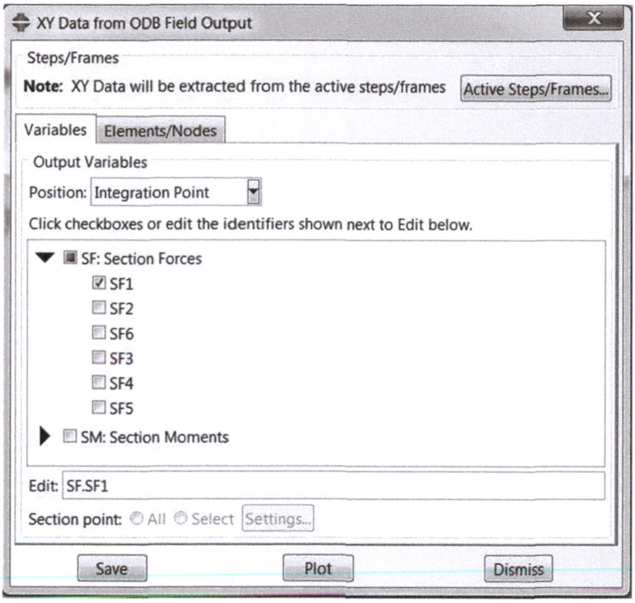

FIGURE 2.14 The axial force SF1 is chosen.

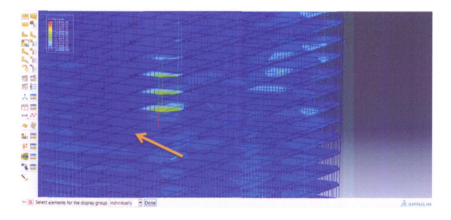

FIGURE 2.15 Choosing a column (highlighted) next to the removed column. (Abaqus® screenshot reprinted with permission from Dassault Systèmes.)

- Go to *Report on the Ribbon* and click on *XY*. A new window will pop up, as shown in Figure 2.16.
- Click on *Setup*, you can define the parameters for the output file.
- Select the *XY* plot you want to output and click *OK*. A text file will be generated that you can copy and paste into an Excel file for further data plotting, as shown in Figure 2.17.

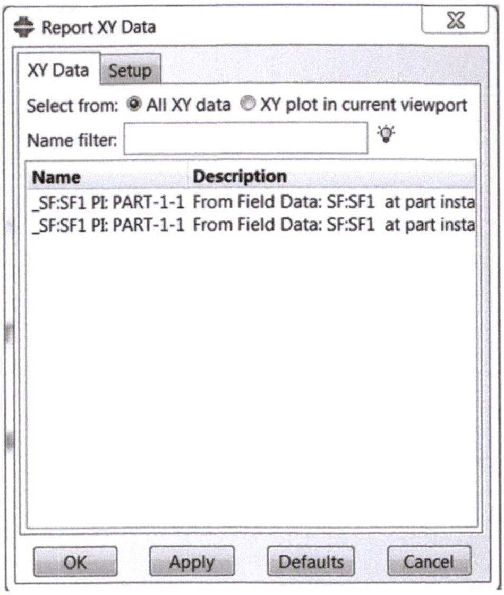

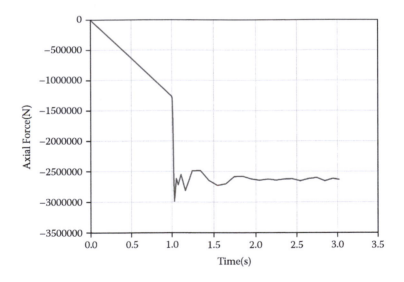

FIGURE 2.16 Exporting axial force to Excel. (Abaqus® screenshot reprinted with permission from Dassault Systèmes.)

FIGURE 2.17 Axial force of the column next to the removed column.

Similarly, readers can also investigate the internal force changing in the beams, as shown in Figure 2.18. A beam (highlighted) above the removed column has been selected.

The internal force, such as the shear force, can be investigated, as shown in Figure 2.19.

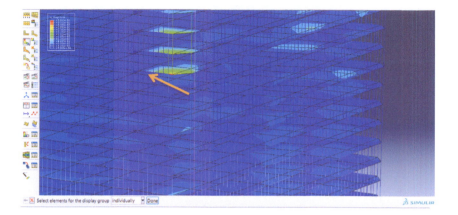

FIGURE 2.18 Choosing a beam (highlighted) above the removed column. (Abaqus® screenshot reprinted with permission from Dassault Systèmes.)

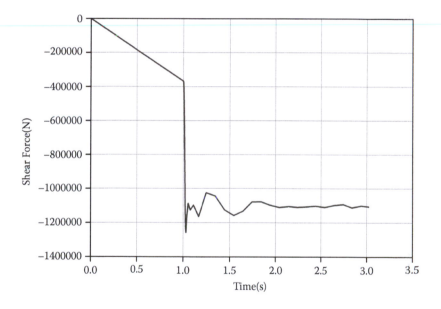

FIGURE 2.19 Shear force development in selected beam.

2.10.6 Progressive Collapse Potential Check

It can be seen from Figure 2.17 that the first 1 second consists of the static steps. At 1 second, the two columns are removed. It can be seen that the axial force of the adjacent column increases dramatically due to the force redistribution. For design purposes, we can use the force obtained from Figure 2.17 to manually check whether the adjacent column can withstand this increased loading. If the column fails, it will need to be removed as well. By continuing to do these column removal procedures, the progressive collapse potential can be checked. If most structural members fail, then it is deemed that a progressive collapse will occur.

References

ASCE (American Society of Civil Engineers). 2005. Minimum design loads for buildings and other structures. SEI/ASCE 7-05. Washington, DC: American Society of Civil Engineers.

BSI (British Standards Institution). 2001. Structural use of steelwork in buildings. Part 1: Code of practice for design—rolled and welded sections. BS 5950. London: BSI.

BSI (British Standards Institution). 2006. Eurocode 1: Action on structures. Part 1-7: General actions: Accidental actions. BS EN 1991-1-7: 2006. London: BSI, September.

BSI (British Standards Institution). 2010. Eurocode—Basis of structural design: Incorporating corrigenda December 2008 and April 2010. BS EN 1990: 2002 + A1: 2005. London: BSI.

Building Seismic Safety Council. 2000. NEHRP recommended provisions for seismic regulations for new buildings and other structures. FEMA 369. Washington, DC: National Institute of Building Sciences.

DHS (Department of Homeland Security). 2003. Reference manual to mitigate potential terrorist attacks against buildings, providing protection to people and buildings. FEMA 426. Washington, DC: DHS, December.

DOD (Department of Defense). 2009. Design of buildings to resist progressive collapse. UFC 4-023-03. Arlington, VA: Department of Defense, July 14.

Elms, D.G. 1992. Risk assessment. In *Engineering Safety*, ed. D. Blockley, 28–46. Berkshire, UK: McGraw-Hill International.

Ettouney, M. 2004. Development of a progressive collapse analysis method and the PROCAT computer program. Counter Terrorism Technology Support Office (CTTSO) Technical Support Working Group (TSWG).

Fu, F. 2009. Progressive collapse analysis of high-rise building with 3-D finite element modelling method. *Journal of Constructional Steel Research*, 65, 1269–1278.

GSA (General Services Administration). 2003. Progressive collapse analysis and design guidelines for new federal office buildings and major modernization projects. Washington, DC: GSA.

Guo, L., Gao, S., and Fu, F. 2015. Structural performance of semi-rigid composite frame under column loss. *Engineering Structures*, 95, 112–116.

Guo, L., Gao, S., Fu, F., and Wang, Y. 2013. Experimental study and numerical analysis of progressive collapse resistance of composite frames. *Journal of Constructional Steel Research*, 89, 236–251.

HM (Her Majesty's) Government. 2013. The Building Regulations 2010: Structure, A3: Disproportionate collapse. Approved Document A, 2004 edition, incorporating 2004, 2010, and 2013 amendments. London: HM Government.

IStructE (Institution of Structural Engineers). 2013. Manual for the systematic risk assessment of high-risk structures against disproportionate collapse. London: IStructE.

Kim, J., and Kim, T. 2009. Assessment of progressive collapse-resisting capacity of steel moment frames. *Journal of Constructional Steel Research*, 65(1), 169–179.

Levy, M., and Salvadori, M. 1992. *Why Buildings Fall Down: How Structures Fail*, 76–83. New York: W.W. Norton.

Ministry of Housing and Local Government. 1968. Report of the inquiry into the collapse of flats at Ronan Point, Canning Town. London: Her Majesty's Stationery Office.

Muhammad, N.S.H., and Thaer, M.S.A. 2012. New building scheme to resist progressive collapse. *Journal of Architectural Engineering*, 18(4), 324–331.

NIST (National Institute of Standards and Technology). 2007. Best practices for reducing the potential for progressive collapse in buildings. Gaithersburg, MD: NIST, Technology Administration, U.S. Department of Commerce.

NIST (National Institute of Standards and Technology) NCSTAR (National Construction Safety Team). 2005. Federal building and fire safety investigation of the World Trade Center disaster. Final report of the National Construction Safety Team on the Collapses of the World Trade Center towers. Gaithersburg, MD: NIST, December.

Pearson, C., and Delatte, N.J. 2005. Ronan Point apartment tower collapse and its effect on building codes. *Journal of Performance of Constructed Facilities*, 19(2), 172–177.

Sasani, M., and Kropelnicki, J. 2008. Progressive collapse analysis of an RC structure, *The Structural Design of Tall and Special Buildings*, 17, 757–771.

Szyniszewski, S. 2009. Dynamic energy based method for progressive collapse analysis structures congress. 1–10.

Tan, S., and Astaneh-Asl, A. 2003. Use of steel cables to prevent progressive collapse of existing buildings. Presented at Proceedings of the Sixth Conference on Tall Buildings in Seismic Regions, Los Angeles, CA, June 4.

CHAPTER 3

Progressive Collapse Design and Analysis of Space Structures

3.1 Introduction

In this chapter, different forms of space structures are introduced, followed by a discussion of collapse accidents around the world and the collapse mechanism of different types of space structures. At the end of this chapter, a progressive collapse analysis example for a double-layer grid space structure is demonstrated using the programs Abaqus® and SAP2000 (CSI, 2013).

3.2 Major Types of Space Structures

Space structures have been widely used in different types of structures, from long-span to mid-span frames and also short enclosures, closed roofs, floors, exterior walls, and canopies. There are several major types of space structures used in current construction projects, such as double-layer grids, latticed shells, membrane structures, and tensegrity structures.

3.2.1 Double-Layer Grids

Double-layer grids are one of the most popular structures used in current construction practice. They consist of top and bottom square grids with nodal joints connected by diagonal struts. Different configurations of the top and bottom layers can make different grid types. This type of construction resembles a pyramid shape. The steel bars are linked together by the joints to form a uniform roof structure.

3.2.2 Latticed Shells

Latticed shells can be built by either a single-layer or a double-layer grid. For a long-span single-layer grid, the connections are normally designed as rigid to provide rigidity. However, for double-layer grid, due to its greater redundancy and indeterminacy, the joints can be designed as pinned.

Domes are one of the commonly used lattice shell structures. There are various types, such as ribbed domes, Schwedler domes, three-way grid domes, lamella domes, and geodesic domes.

A barrel vault is another latticed shell structure. It features forms of a cylindrical-shaped shell, which has a surface that can be easily modified due to its zero curvature. The location and type of supports between the members also influence the vault's structural behaviour.

3.2.3 Tensegrity Systems

As shown in Figure 3.1, tensegrity structures are self-equilibrium systems composed of continuous prestressed cables and individual compression bars. They are one of the most promising solutions for large-span space structures due to their superlight weight. The concept of tensegrity was first conceived by R.B. Fuller (1975), reflecting his idea of "nature relies on continuous tension to embrace islanded compression elements." D.H. Geiger et al. (1986) made use of Fuller's thought and designed an innovative "cable dome" in the circular roof structures of the gymnastic and fencing arenas (Figures 3.1 and 3.2) for the Seoul Olympic Games (Geiger et al., 1986).

FIGURE 3.1 Interior of the fencing arenas for the Seoul Olympic Games. (Photo taken by the author.)

FIGURE 3.2 Fencing arenas for the Seoul Olympic Games. (Photo taken by the author.)

3.2.4 Membrane Structures

Membrane structures are a type of lightweight space structure. The membrane works together with lattice shell or tensegrity structures (such as the gymnastic and fencing arenas for the Seoul Olympic Games; Geiger et al., 1986), as shown in Figure 3.2. Some are made as inflatable structures.

3.3 Design Guidance for Space Structures to Prevent Disproportionate Collapse

Space structures consist of a large number of structural members. For most types of space structures, due to their structural redundancy, most designers presume that a progressive collapse will not be triggered when the loss of an individual member occurs. However, as it will be introduced in Section 3.4, there have still been a number of space structure collapse incidents that have been reported worldwide.

So far, there are few design codes with detailed requirements for progressive collapse prevention designs of space structures, although several past codes of practice provide design procedures for very long bridges.

3.4 Space Structure Collapse Incidents around the World

In this section, an introduction of collapse incidents of space structures around the world is made. The cause of the collapse and the failure mechanism of each incident are introduced.

3.4.1 Partial Collapse of Charles de Gaulle Airport Terminal 2E

On May 23, 2004, a portion of Terminal 2E's ceiling collapsed near Gate E50. The structure was designed with a 300 mm thick curved concrete shell with a span of 26.2 m, which was precast in three parts, as shown in Figure 3.3.

The two sides of the structure were externally strengthened with curved steel tension members, which were propped with struts. The shell rested on two longitudinal support beams that were supported and tied back to the columns. At the location of the failure, there were large openings for access to gangways. The external steelwork and shell were enclosed with glass.

The investigation report of the Ministry of Transportation (Conseil National des Ingenieurs et Scientifiques de France, 2005) found a number of causes for the collapse. The main reason was that the steel dowels supporting the concrete shell were too deeply embedded into it, which caused cracking in the concrete. In turn, this led to a weakening of the roof. The cracks were formed due to high stress caused in

FIGURE 3.3 Partial collapse of Charles de Gaulle Airport Terminal 2E. (From https://upload.wikimedia.org/wikipedia/commons/1/1b/Terminal_2E_CDG_collapse.png. Free licence.)

the construction stage and cycles of stress from differential thermal and moisture movements.

The investigation also found that the structure had little margin for safety in design, and a combination of factors led to the major collapse, including

- High flexibility in the structure under dead load and external actions
- Cracking that may have resulted from insufficient or misplaced reinforcement
- Lack of robustness and redundancy to transfer loads away from a local failure
- High local punching stresses where the struts were seated in the concrete shell
- Weakness of the longitudinal support beam and its horizontal ties to the columns

The above design fault caused progressive collapse between the concrete shell and curved steel tension member and struts.

3.4.2 Snow-Induced Collapse of Double-Layer Grid Space Structure, Hartford Civic Center

Heavy snow is another major cause for the collapse of space structures. O'Rourke and Wikoff (2014) described an investigation into about 500 roof collapse incidents that occurred in the northeastern United States during the winter of 2010–2011. The major reason for these collapses was the snow load exceeded the design load required by the building code or the structural member was designed with a structural capacity that was significantly less than that required by the building code.

One famous collapse incident is that of the Hartford Civic Center Coliseum in 1978, due to the largest snowstorm in a 5-year time period. The snow loading caused excessive deflection of the space frame roof, which casuses the final collapse.

Three independent investigations have been done. The space frame construction for the stadium was a double-layer grid. In the conventional design of double-layer grid structures, the centrelines of each member intersect into the same joint to reduce the bending moment. However, the investigation report (Lev Zetlin Associates, 1978) shows that in the case of the Hartford Civic Center's frame, the top chords intersected at one point and the diagonals at another, which caused bending stresses in the members. In addition, the lateral bracing of the top chords was met through diagonals in the interior of the frame, but along the edges there was no means to prevent out-of-plane bending (Lev Zetlin Associates, 1978).

A faulty weld connecting the scoreboard to the roof was also noticed. A massive amount of energy would have been caused by the volatile weld release, causing the entire structure to collapse (Feld and Carper, 1997).

It was also noticed that once the roof truss was in place, the construction manager altered the roof material, increasing the dead load by 20% (Feld and Carper, 1997). Therefore, the dead loads were underestimated by more than 20%.

Another investigation showed that the cause of the failure was due to torsional buckling of the compression members, and that members close to the middle of the truss were critically loaded even before live loads were added. This means of failure is usually overlooked as a cause of failure because it is so uncommon (ENR, 1978).

3.4.3 Roof Collapse of Pavilion Constructed in Bucharest

Another collapse example is a pavilion constructed in Bucharest in 1963 (Vlad and Vlad, 2014). The pavilion was a braced dome with a span of 100 m and rise of 0.48 m. The dome collapsed as a result of local snap-through due to an unexpected snow load accumulation on a small area. The local buckling propagated rapidly, and this propagation of deformation caused the dome to collapse.

3.4.4 Roof Collapse of Sultan Mizan Stadium in Terengganu, Malaysia (Support Failure)

The roof of the football stadium in Terengganu, Malaysia, was constructed as a curved double-layer grid. It collapsed 1 year after completion, in 2009. The primary cause for the collapse was incomplete consideration of the support conditions for the roof.

The report from the investigation committee explains the reason for the collapse (Investigation Committee on the Roof Collapse at Stadium Sultan Mizan Zainal Abidin, 2009):

- The design was inadequate; the designer failed to fully take into account the support conditions of the roof structure.
- The complexity and long spans of the roof structure required more detailed consideration in second-order design analysis, which was not carried out.
- The sensitivity of the space frame roof structure required consideration of the support flexibility in the design mode, which was not done.

In the construction stage, the roof was erected poorly, resulting in misaligned geometry; poor workmanship was another reason for collapse.

On February 20, 2013, the stadium collapsed again while undergoing reconstruction work. Two-thirds of the old structure (137 m) collapsed, followed by the collapse of steel pillars. The collapse was due to the removal of the middle framework.

From these two collapses in one stadium, it can be seen that the support failure was one of the major reasons for the collapse of the space structure. A modelling analysis considering the support failure is demonstrated later in this chapter.

3.5 Collapse Mechanism of Space Structures

As introduced in Section 3.1, there are different types of space structures. Therefore, the collapse mechanism varies depending on their structural form. In this section, the collapse mechanism for different types of space structures will be explained.

3.5.1 Collapse Mechanism for Double-Layer Grid

A double-layer grid space structure is one of the conventional long-span structures. Due to its large statical indeterminacy and redundancy of structural members, in design practice, it is normally considered that progressive collapse will not be triggered when the loss of an individual member occurs. However, in the research presented by Murtha-Smith (1988), an analysis was performed on hypothetical space trusses and showed that progressive collapse could occur following the loss of just one of several potentially critical members when the structures were subject to full service loading.

The collapse of the Hartford Coliseum showed that progressive collapse could also occur following the loss of some critical members when the structures are overloaded by a gravity load, such as excessive snow loading due to severe weather.

In addition to snow, strong wind was also found to be a reason for the collapse of a building; therefore, in real design practice, a wind tunnel test is normally required for long-span space structure designs.

Based on the above discussion, in the design, an extra safety margin should be made for structural members to resist extra loading, therefore to prevent progressive collapse due to abnormal gravity loads.

3.5.2 Collapse Mechanism of Single-Layer Space Structures

For certain types of space structures, such as the single-layer lattice shell, stability is important, as buckling may initiate the collapse of the structure. Structures such as single-layer braced domes are prone to progressive collapse due to propagation of local instability initiated by member or node instability. Research from Abedi and Parke (1996) found that, for single-layer braced domes, the dynamic snap-through is associated with inertial effects and large localized deformation and can propagate to lead to collapse.

The reason is that for all the space structures introduced in this chapter, a single-layer space frame exhibits greater sensitivity to buckling than a double-layer structure. It was also found that a shallow shell, such as a dome, is more prone to overall buckling than a cylindrical shell. And global buckling may be triggered if certain critical members fail. An example is a roof collapse of the pavilion constructed in Bucharest.

To help readers fully understand the buckling behaviour of the lattice shell, it is worthwhile to introduce the major types of buckling here. In design practice, there are three major types of buckling that need to be checked: member buckling, local buckling of certain members, and global buckling.

Member buckling is when the individual member becomes unstable; it includes the overall buckling of the members and local buckling of the flanges or webs. The corresponding design formulas are given in worldwide guidances such as Eurocode 3 (European Committee for Standardization, 2005). The theories behind these design formulas are Euler's buckling theory and the Perry–Robertson equations. A designer can check the stability of individual structural members accordingly.

Local buckling consists of a snap-through buckling of a group of structural members in a local area, which often takes place at joints. The local buckling of a space frame is prone to happen in single-layer structures such as single-layer lattice shells. The type of buckling collapse of a space frame is greatly influenced by the curvature and thickness of the structure and the manner of supporting and loading. It is apt to occur when t/R is small, where t is the thickness of the structure and R is the radius of curvature.

Global buckling refers to a relatively large area of the space frame becoming unstable. It is often triggered by local buckling. Therefore, global buckling analysis of the whole structure should also be performed in the design to prevent progressive collapse.

3.5.3 Collapse Mechanism of Tensegrity Structures

Different than conventional space structures, tensegrity structures have a unique feature in their collapse mechanism due to their self-equilibrium system. Research from Kahla and Moussa (2002), Abedi and Shekastehband (2008), and Shekastehband et al. (2011) shows that the behaviour of members has a dominant effect on the overall collapse behaviour of space structures.

In tensegrity structures, a member may suddenly fail in tension (cables) or compression (struts). These may be due to the snap-through of struts under compression and cable ruptures under tension. Similar to single-layer domes, an initial failure of a small portion of the structure has the potential to propagate to other parts of the structure and may ultimately cause overall collapse.

In fact, member failure has a dynamic effect on the behaviours of the whole system, as it releases kinetic energy in a local region of the structure. Therefore, it is important to account for dynamic effects in the analysis, especially the redistribution of member forces and inertia forces caused by the member failure, when evaluating the response of these structures under the member failure phenomenon (Shekastehband et al., 2012; Shekastehband and Abedi, 2013).

3.5.4 Support Failure

From the incidents introduced in Section 3.2, it can be seen that support failure is one of the key reasons for the collapse of all types of space structures. Improper construction methods (such as in Mizan Stadium in Terengganu), heavy earthquakes, and foundation settlement will all cause support failure. Therefore, in the design, an engineer should be able to check the support collapse potential of the space structures with support failure. Detailed analysis using Abaqus® is shown in Section 3.6.

3.6 Progressive Collapse Analysis of Double-Layer Grid Space Structure Using Abaqus®

As mentioned in Section 3.4.2, an excessive gravity load may cause the collapse of a double-layer grid space structure. Therefore, in this section, how to perform a progressive collapse analysis of the double-layer grid space structure is demonstrated using the commercial programs Abaqus® and SAP2000 (CSI, 2013).

3.6.1 *Prototype Space Structure*

In research by Fu and Parke (2015), a double-layer grid space structure was modelled as the prototype. It was a conventional square grid, 27 m long on each side, and consisted of 324 square pyramids.

These kinds of systems are sometimes formed by continuous top and bottom chords, with pinned diagonal struts forming the web members. However, the normal construction of these structural types is to use individual tubular members, spanning from node to node for the top and bottom chords, and additional tubular members for the diagonal web members. All of the members are generally considered to be pinned. Therefore, in the simulation, pin connections are modelled for all the members.

The height of the grid was 1.5 m. The whole structure was vertically supported at selected perimeter nodes in the locations shown in Figure 3.4. The support-to-support span was 9 m, giving a span-to-depth ratio of 6.

3.6.2 *Setting Up a 3D Model*

The geometry of the space structure is sophisticated. It is more efficient to set the three-dimensional (3D) model in software such as Rhino or AutoCAD, and then import it into an analysis program such as SAP2000 or Abaqus®. In this analysis, a 3D model was set up first in SAP2000, as shown in Figure 3.4. This is because

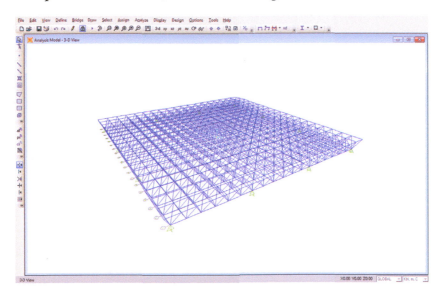

FIGURE 3.4 Double-layer grid model in SAP2000. (SAP2000 screenshot reprinted with permission of CSI.)

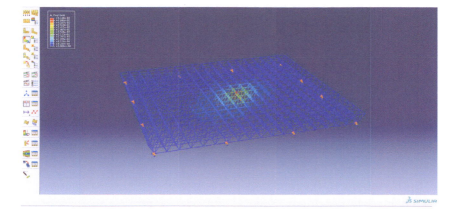

FIGURE 3.5 Double-layer grid model in Abaqus® (a middle member is removed). (Abaqus® screenshot reprinted with permission from Dassault Systèmes.)

SAP2000 is a design-oriented program with design code incorporated. Therefore, the model was first analysed and designed under normal loading conditions using SAP2000 to make sure no member was overstressed and that no overall buckling and local buckling of the structural members were observed. To simplify the analysis and design, 60.3 × 5 CHHFs—60 mm diameter pipe sections with a 5 mm wall thickness—were chosen for all the members. The yield stress of the chosen steel was 355 N/mm².

As shown in Figure 3.5, after analysis in SAP2000, the 3D model can be imported into Abaqus® CAE. However, similar to Chapter 2, the analysis is conducted using INP files here.

In the proposed model, all of the top and bottom chord and diagonal members were modelled using *BEAM elements. The material properties of all the structural steel components were modelled using elastic–plastic material behaviour from Abaqus® incorporating material nonlinearity. The elastic part of the stress–strain curve was defined with the *ELASTIC option, and the values 2.06 × 105 N/mm² for Young's modulus and 0.3 for Poisson's ratio were used. The plastic part of the stress–strain curve was defined with the *PLASTIC option. Steel grade S355 was used for all the structural steel. Engineering stresses and strains, including the yield and ultimate strength, were obtained from BS 5990 (BSI, 2001) and converted into true stresses and strains with the appropriate input format for Abaqus®.

3.6.3 Load Combinations

For nonlinear dynamic analysis, the GSA guideline (2003) has the load combination requirement of dead load plus 0.25 of the live load.

However, with reference to the collapse incident of the Hartford Civic Center, to make the analysis more conservative, the full live load was used, so the load combination used in the analysis was 1.0 dead + 1.0 live (live load is taken as 1 kN/m² in the analysis).

3.6.4 Major Abaqus® Command Used in the Simulation
Similar to Chapter 2, the INP file consists of several main parts. Readers can refer to Chapter 2 for detailed examples.

3.6.5 Member Removals
The members to be removed were forcibly removed by instantaneously deleting them. Several removal scenarios were selected; they are shown in Table 3.1.

3.6.5.1 Case 1   In the first analysis, as shown in Figure 3.5, one web member, which was at the centre of the grid, was removed. However, as shown in Figure 3.6, no obvious dynamic response was observed.

Table 3.1 Member Removal Scenarios

Case 1	Removal of a structural member at centre
Case 2	Removal of a square pyramid at centre
Case 3	Centre support failure

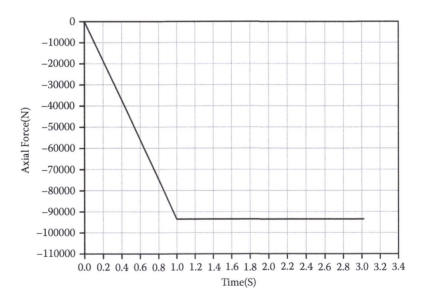

FIGURE 3.6 Response of axial force in the top chord near the removed member for case 1.

It can be seen that due to the redundancy of the structure, a single member failure does not have a significant effect on the structure.

3.6.5.2 *Case 2* In case 2, all of the web members in one square pyramid, located at the centre of the structure, were removed. Figure 3.7 shows the contour plot of vertical displacement after the central pyramid removal. The response of the axial force in the top chord near the removed central pyramid is shown in Figure 3.8. A dynamic response was observed. It should be noted that the first second consists of the static step; the static load (live + dead) was applied in this analysis step, and the axial force increased from 0 to the maximum force. After the first second, the dynamic procedure started, whereupon the structural members were removed. The response of the axial force of a diagonal strut is shown in Figure 3.9.

A design check was made after the analysis. The tensile capacity of each structural member was 308 kN, the buckling load for the top and bottom chords was 215 kN, and the buckling load for the diagonal struts was 144 kN. This indicated that no further member failure occurred after the removal of the square central pyramid.

3.6.5.3 *Case 3* In this analysis, in the middle of one edge, support A was removed. The removal was done by deleting several members connected to the support. This was also to simulate the support failures as they occurred in Sultan Mizan Stadium. The removed

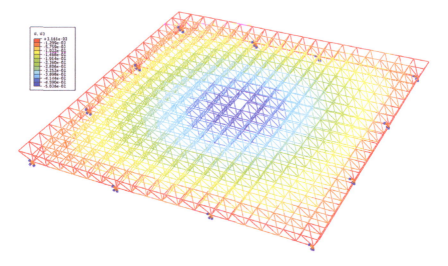

FIGURE 3.7 Contour plot of vertical displacement (one central pyramid removal). (Abaqus® screenshot reprinted with permission from Dassault Systèmes.)

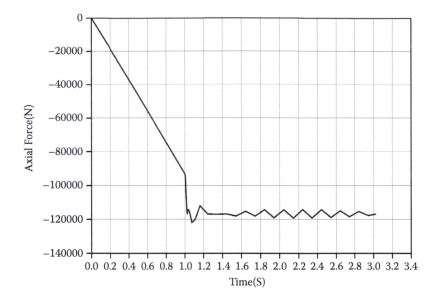

FIGURE 3.8 Response of axial force in the Top chord near the removed central pyramid (case 2). (Abaqus® screenshot reprinted with permission from Dassault Systèmes.)

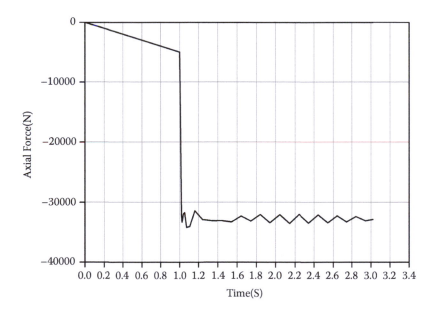

FIGURE 3.9 Response of axial force in the diagonal strut near the removed central pyramid (case 2). (Abaqus® screenshot reprinted with permission from Dassault Systèmes.)

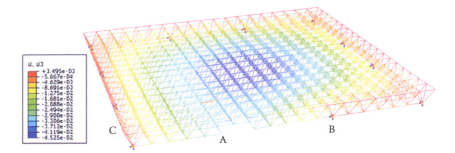

FIGURE 3.10 Contour plot of vertical displacement after removal of central support A. (Abaqus® screenshot reprinted with permission from Dassault Systèmes.)

members and the location of support A are shown in Figure 3.10, which also shows the distribution of the vertical deflection in the structure after the removal of support.

Figure 3.11 shows the stress contour after removal of the support. It can be seen that after removal of support A, some structural members close to the adjacent support (B) become overstressed. This is because most of the loads carried by support A were redistributed into the remaining support, primarily those supports at locations B and caused the overstresses in the members close to support B.

Figure 3.12 shows the axial force in the bottom chord near support B. It can be seen that after removal of the support, the bottom

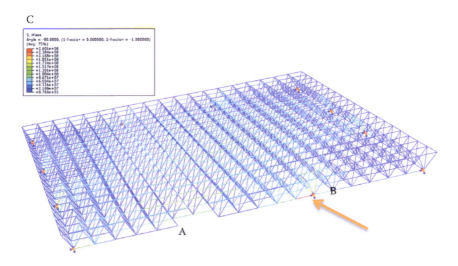

FIGURE 3.11 Contour plot of stress. (Abaqus® screenshot reprinted with permission from Dassault Systèmes.)

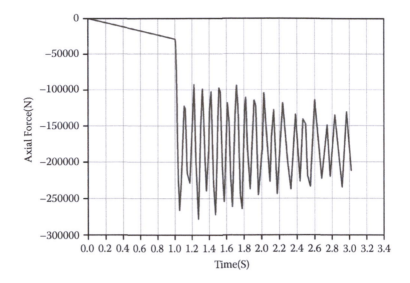

FIGURE 3.12 Axial force in the bottom chord near support B.

chord buckled as the axial force exceeded the buckling capacity, which is 215 kN for the relevant members.

3.6.5.4 Progressive Collapse Potential Check Based on the above analysis, we can also note that due to the high redundancy of the structure—for a space grid, supporting a normal live load—removal of a single structural member is unlikely to trigger the collapse of the whole structure. However, under an abnormal live load condition, such as a very heavy snow load, a progressive collapse of the structure can be triggered, as it may cause the failure of several members, which will increase the possibility of a progressive collapse.

However, great attention also needs to be paid to support failures, because when a support fails, the load is redistributed to the adjacent supports, which are likely to cause further member failures and trigger a progressive collapse. Detailed analysis results can be found in Fu and Parke (2015).

References

Abedi, K., and Parke, G. 1996. Progressive collapse of single-layer braced domes. *International Journal of Space Structures*, 11(3), 291–305.

Abedi, K., and Shekastehband, B. 2008. Static stability behavior of plane double-layer tensegrity structures. *International Journal of Space Structures*, 23, 89–102.

BSI (British Standards Institution). 2001. Structural use of steelwork in buildings. Part 1: Code of practice for design—rolled and welded sections. BS 5950. London: BSI.

Conseil National des Ingenieurs et Scientifiques de France. 2005. Synthese Des Travaux de la commission administrative sur les causes de l'eddondrement d'une partie du terminal 2E de Paris-Charles de Gaulle. Paris: Ministry of Transportation, Urban Design, Tourism, and Sea.

CSI (Computer and Structures). 2013. *SAP2000 theory manual.* New York: Computer and Structures.

ENR. 1978. Collapsed space truss roof had a combination of flaws. *ENR*, June 22.

European Committee for Standardization. 2005. Eurocode 3: Design of steel structures. Part 1-1: General rules and rules for buildings. BS EN 1993-1-1. London: European Committee for Standardization.

Feld, J., and Carper, K.L. 1997. *Construction Failures*, 198. 2nd ed. New York: Wiley & Sons.

Fu, F., and Parke, G. 2015. Assessment of the progressive collapse resistance of double-layer grid space structures. Under review.

Fuller, R.B. 1975. *Synergetics Explorations in the Geometry of Thinking.* London: Collier Macmillan Publishers.

Geiger, D., Stefaniuk, A., and Chen, D. 1986. The design and construction of two cable domes for the Korean Olympics: Shells, membranes and space frames. In *Proceedings of IASS Symposium*, Osaka, vol. 2, pp. 265–272.

GSA (General Services Administration). 2003. Progressive collapse analysis and design guidelines for new federal office buildings and major modernization projects. Washington, DC: GSA.

Investigation Committee on the Roof Collapse at Stadium Sultan Mizan Zainal Abidin. 2009. Final report on the roof collapse at Stadium Sultan Mizan Zainal Abidin, Kuala Terengganu, Terengganu Darul Iman. Vols. 1–3, December.

Kahla, N.B., and Moussa, B. 2002. Effect of a cable rupture on tensegrity systems. *International Journal of Space Structures*, 17, 51–65.

Lev Zetlin Associates. 1978. Report of the engineering investigation concerning the causes of the collapse of the Hartford Coliseum space truss roof on January 18, 1978. June 12.

Murtha-Smith, E. 1988. Alternate path analysis of space trusses for progressive collapse. *Journal of Structural Engineering*, 114(9), 1978–1999.

O'Rourke, M., and Wikoff, J. 2014. *Snow-related roof collapse during the winter of 2010–2011: Implications for building codes.* Washington, DC: American Society of Civil Engineers.

Shekastehband, B., and Abedi, K. 2013. Collapse behavior of tensegrity systems due to cable rupture. *International Journal of Structural Stability and Dynamics*, 13, 1250079.

Shekastehband, B., Abedi, K., and Chenaghlou, M.R. 2011. Sensitivity analysis of tensegrity systems due to member loss. *Journal of Constructional Steel Research*, 67, 1325–1340.

FIGURE 4.1 Collapse of the Kutai Kartanegara Bridge (From https://commons
.wikimedia.org/wiki/File:Sisi_utara_jembatan.jpg. Free licence by Wiki.)

anchorages should be designed to hold the cables of the bridge.
Suspension bridges are often vulnerable to collapse when two or
three hangers fail, which can cause a progressive collapse as the
remaining supports become overloaded.

The truss for the main span of the Kutai Kartanegara Bridge
was 635 m long; it collapsed into the Mahakam River in the East
Kalimantan Province. The investigation of Lynch (2011) shows that
the collapse was due to hanger maintenance work. The site workers were adjusting the tension of the bridge hangers just before the
collapse. A progressive collapse was triggered from the failure of
one or more of the hangers during the maintenance. The connections between the hangers and truss were subject to fatigue (caused
by traffic) and corrosion during their service period. Therefore,
they required close monitoring and special maintenance. However,
due to their structural form, it was difficult to access the hangers for maintenance, making the work more difficult. The investigation report by Indonesia's Ministry of Public Works (Lynch,
2012) shows that failure occurred when the engineers were jack-
ing underneath the bridge deck, which caused extra stress on the
hanger connection. This caused shear failure of the bolt connection between the steel hanger and bridge deck. After one connection failed, the increased stress on the other connections caused
a chain reaction, resulting in the bolts failing in the remaining
connections and failure of the remaining hangers.

4.2.2 Collapse of Skagit River Bridge, Washington, Due to Lorry Strike

In 2013, a span of the Skagit River Bridge in the United States collapsed into the river (Figure 4.2). The cause of the collapse was an oversize truck striking several overhead support beams of the bridge, which led to an immediate collapse of the northernmost span. The progressive collapse of one span of the bridge was due to the bridge's design; it did not have redundant structural members to protect its structural integrity in the event of a failure of one of the bridge's support members.

Structurally, the bridge consists of four consecutive independent spans. Therefore, the failure of one span of the bridge did not trigger progressive collapse of the whole bridge. Only the deck and overhead superstructure of the northernmost span collapsed into the river; the piers below the deck were not damaged.

4.2.3 Earthquake-Induced Collapse of Bridge in Wenchuan, China

One of the major reasons for the collapse of the bridge was the failure of the piers, caused by a major earthquake. The Wenchuan earthquake occurred in Sichuan Province, China, on May 12, 2008, with a magnitude of 8.0. Several bridges collapsed during this earthquake.

BaiHua Bridge is a 500 m long post-tension concrete girder bridge. Four spans of the bridge collapsed. The structural form of the bridge

FIGURE 4.2 Skagit River Bridge collapse. (From https://commons.wikimedia.org/wiki/File:05-23-13_Skagit_Bridge_Collapse.jpg. Free to use by Wikipedia.)

is a moment resisting frame consisting of piers and lateral beams. Other parts of the bridge did not collapse.

The investigation of Kawashima et al. (2009) shows that for these collapsed spans, due to the strong ground motion, the lateral beam was detached and the pier was broken into two pieces. The study also shows that for some of the lateral beam–column joints, the surface of the detached lateral beam was very smooth, which apparently demonstrates that plastic hinges were not formed. In some joints, only 10 mm diameter hoops were provided at 300 mm intervals, resulting in poor lateral confinement of the joints.

Therefore, the insufficient amounts of reinforcement resulted in a brittle failure at the joints during the earthquake, which was the major reason for the collapse of the bridge.

4.3 Causes and Collapse Mechanisms of Bridges

Compared to buildings, progressive collapse has not been a major consideration for bridge structures in the past. However, the collapse incidents introduced in Section 4.2 show the importance of the design of a bridge structure to prevent progressive collapse. Therefore, as a priority, it is imperative to understand the causes and collapse mechanisms of bridges. It can be seen from Section 4.2 that the collapse mechanisms of bridges are heavily dependent on the structural types of the bridges. In this section, the major causes of collapse and the correspondent collapse mechanisms of different types of bridge structures are introduced.

4.3.1 Accidental Actions (Impact Loads) Triggered Collapses

An accidental action, such as a ship collision, is one of the major causes of bridge collapse.

There are several scenarios of impact cases, such as a boat collision with a bridge pillar or side collision or ship deckhouse collision with a bridge span.

The factors that affect the impact load on a bridge are the type of waterway, the flood conditions, the type and draught of vessels, and the type of structure (JRC-Ispra, 2012).

There are several methods to calculate the impact loading. It is worthwhile to introduce them here for the readers' reference.

The AASHTO specifications (2007) provide the below formula:

$$F = 1.11 \cdot 0.88\sqrt{DWT} \cdot \frac{V}{8}$$

where
> F is the equivalent average impact load in MN
> v is the ship impact speed in m/s.

Eurocode 1, Part 2-7 (European Committee for Standardization, 2003), provides another method to calculate the design collision:

$$F = \sqrt{Km} \cdot V$$

where
> K is an equivalent stiffness,
> M is the impact mass, and
> V is the velocity.

This Eurocode equation gives slightly higher values than the AASHTO equation. The reader can make choice according to the specific projects you are working on.

One of the conventional methods to prevent accidental action–triggered collapse is to build protective structures, such as artificial islands or guide structures, to protect the bridge pier. Another method is to design piers to be strong enough to withstand direct collisions. However, both methods increase the overall cost of projects.

4.3.2 Earthquake-Induced Collapse of Bridge

As discussed in Section 4.2, earthquake is another major reason for bridge collapse. Historically, bridges have been vulnerable to earthquakes. Earthquake cause damage to substructures and foundations and, in some cases, the collapse of the entire bridge.

In addition to the Wenchuan earthquake, there have been a number of earthquake-induced collapses or partial collapses of bridges around the world. In 1964, nearly every bridge along Cooper River Highway in Alaska was seriously damaged or destroyed due to a magnitude 9 earthquake. In 1971, more than 60 bridges were damaged in the San Fernando earthquake. In 1989, more than 80 bridges were damaged in the Loma Prieta earthquake in California.

There are several failure modes that can be identified for bridge structures under earthquake loading, such as flexural failure at the base of concrete piers, shear failure of concrete piers, unseating of the simply supported link span, bond failure of lap slices in concrete bridge piers, and the combined flexural–shear failure of concrete piers. Among those incidents, bridge pier failure is one of the major reasons for bridge collapse.

Because flexural and shear failures are the most common failure modes for concrete piers, they are summarized here.

4.3.2.1 Flexural Failure of Piers Due to the large horizontal motion, flexural failure of piers is one of major failure modes. The flexural failure at the base of the bridge pier of the Hanshin Expressway during the 1995 Kobe earthquake was the major cause for the collapse of the bridge. A similar failure mode was observed during the Wenchuan earthquake for concrete bridges. Flexural failure is more predictable than shear failure. In the design process, designers should provide sufficient longitudinal reinforcement with enough anchorage and lapping to ensure sufficient ductility can be developed in the pier. The investigation of bridge collapses during the Wenchuan earthquake, mentioned in Section 4.2.3, also shows that hoops should be designed to provide sufficient lateral confinement of the joints; therefore, plastic hinges can be formed and brittle failure avoided. As it is widely known, plasticity helps to dissipate energy input from earthquakes.

4.3.2.2 Shear Failure of Piers Depending on the type of pier-to-deck connections, the type of foundation, the height of the pier, and the reinforcement detailing of the pier, shear failure can also be observed, as shown in Figure 4.3. The shear failure for a concrete

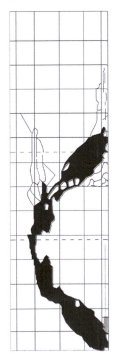

FIGURE 4.3 Shear failure of bridge pier.

FIGURE 4.4 Tacoma Narrows Bridge collapse. (From https://upload.wikimedia .org/wikipedia/commons/4/4a/Tacoma-narrows-bridge-collapse.jpg. Public domain confirmed from Wikimedia.)

pier is mainly caused by the failure of transverse reinforcements. Because shear failure is sudden and brittle, so this type of failure should be prevented in design. Engineering experience tells us that short columns are prone to shear failure; therefore, they should also be avoided in the design of bridge piers.

4.3.3 Wind-Induced Collapse of Bridge

For certain types of bridge structures, especially suspension bridges, strong wind is another major cause for bridge collapse. As shown in Figure 4.4, the Tacoma Narrows Bridge collapse is a famous example for one due to strong wind. On November 7, 1940, Tacoma Narrows Bridge collapsed in a 42 mph (68 km/h) gust (Billah and Scanlan, 1991). The video documentary of the failure of the Tacoma Narrows Bridge demonstrates the warbling of the deck, where wind-induced damage to the superstructure resulted in the loss of nearly the entire main span.

The major reason for the collapse was the aeroelastic flutter caused by the strong wind. Its failure boosted research in the field of bridge aerodynamics–aeroelastics, the study of which has influenced the designs of all the world's great long-span bridges built since 1940.

4.4 Design Measures to Prevent Bridge Collapse

As introduced in other chapters, there are a number of design guidances on how to prevent progressive collapse. However, so far, there are few clear guidances on how to design a bridge to prevent progressive collapse. Progressive collapse has not been a major consideration for bridge structures. Cable-stayed bridges are the only type of bridges with a requirement to check the cable loss (PTI, 2007). The Post-Tensioning Institute (PTI, 2007) states that some special requirement is needed to make sure progressive collapse is not triggered. Both the International Federation for Structural Concrete (FIB, 2005) and PTI (2001) require that the loss of any one cable should not lead to structural instability.

As introduced in the earlier sections, the collapse mechanisms for bridge structures are structure type dependent; therefore, for different structural types of bridges, different strategies to prevent their collapse should be adopted. A detailed introduction is made here.

4.4.1 Beam Bridge

For a beam bridge or continuous beam bridge, the pier is one of the key elements to be designed. In the design, the initial local failure of the pier should be prevented by increasing the level of safety or design to resist abnormal loadings. In addition, due to the importance of the bridge, a cofferdam or buffer zone can be designed to prevent the pier from abnormal impact loadings, such as ship collision.

Another method is the so-called segmentation design method (Starossek, 2006), where a bridge consists of consecutive independent spans. Therefore, if one span fails by accident, the remaining span will not be affected and will be kept intact.

4.4.2 Cable-Stayed Bridge

A cable-stayed bridge features a high degree of static indeterminacy and redundancy. However, progressive collapse cannot be overlooked. It is evident that the cable is the key element in collapse design. The loss of cables can lead to overloading and the rupture of adjacent cables. Therefore, PTI (2001) and FIB (2005) require that a cable-stayed bridge be capable of withstanding the loss of any one cable without the occurrence of structural instability.

The abrupt loss of a single cable will govern the design. However, as required by PTI (2001), the potential for multiadjacent cable loss should also be checked due to terrorism or an accident.

4.4.3 Suspension Bridge

As mentioned in the Kutai Kartanegara Bridge collapse example, for suspension bridges, the hanger is one of the important load-resistant elements. The sudden failure of a hanger will lead to an impulsive dynamic loading on the remaining system and a slackening of the adjacent hangers. Therefore, a collapse of the bridge may be triggered.

As found by Zoli and Steinhouse, this happens because the hangers are connected to deep stiffening trusses; they are not able to efficiently redistribute localized loading due to the member loss of a structure. Therefore, the force will redistribute to adjacent hangers, which may result in their failure. Thus, designers cannot simply treat the superstructure as a fuse to protect adjacent hangers. Large margins of safety in the design of hangers are recommended to prevent progressive collapse.

It is obvious that another load-bearing element of a suspension bridge is the suspension cable. As stated by Starossek (2006), increasing the section area of cable could be an appropriate measure for small spans. For large-span bridges, this is not practical due to the costs. However, large-span bridges exhibit a huge cross-sectional area and mass that prevent local failure without further strengthening.

4.5 Progressive Collapse Analysis of Bridge Structures

In Chapter 2, different progressive analysis procedures for building structures were introduced. However, little guidance is provided on how to conduct a progressive analysis for bridge structures. PTI (2001) provides prescriptive guidance in the extreme event of cable loss, in terms of load applications and resistance factors. Two load application methods are prescribed. The simplified static method is to investigate the structure with a missing cable under factored dead and live loads, combined with the static application of the dynamic force imparted from the severed cable. Alternatively, PTI (2001) permits the use of dynamic procedures to determine the response of the bridge due to cable loss. However, a detailed analysis procedure is not explained. As it is widely recognized (Fu, 2009) that nonlinear dynamic analysis is one of the most accurate procedures, it is introduced here for progressive analysis of bridge structures.

4.6 Progressive Collapse Analysis Example of the Millau Viaduct Using Abaqus® (Nonlinear Dynamic Procedure)

In order to demonstrate the way to perform progressive collapse analysis, as shown in Figure 4.5, the Millau Viaduct designed by

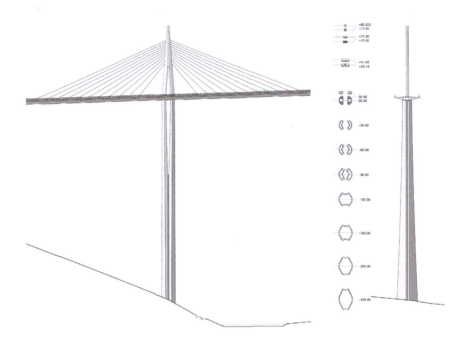

FIGURE 4.5 Millau Viaduct (Architectural drawings provided by Foster + Partners. Image courtesy of Foster + Partners.)

Foster + Partners is used as a prototype. The 3D bridge model was set up using Abaqus® based on the architectural drawings provided by Foster + Partners (Figure 4.6). A nonlinear dynamic procedure for progressive collapse analysis was performed; detailed modelling procedures are demonstrated in the following sections.

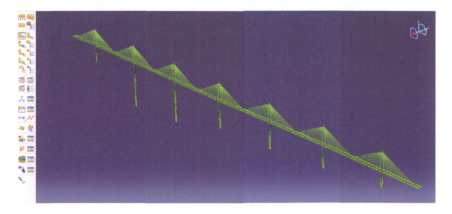

FIGURE 4.6 3D model of Millau Bridge in Abaqus®. (Abaqus® screenshot reprinted with permission from Dassault Systèmes.)

4.6.1 Introduction of Prototype Bridge

Millau Viaduct is an eight-span cable-stayed structure with a complete length of 2460 m; it is currently the tallest bridge in the world. The six central spans have a length of 342 m, and the two end spans have a length of 204 m each. The seven piers with different height are shown in Table 4.1.

The deck of Millau Bridge is in the form of a trapezoidal profiled steel girder box with a depth of 4.20 m and an orthotropic decking made up of metal sheets (Figure 4.7). As shown in Figure 4.5, the

Table 4.1 Height of the Piers

Pier 1	Pier 2	Pier 3	Pier 4	Pier 5	Pier 6	Pier 7
94.50 m	244.96 m	221.05 m	144.21 m	136.42 m	111.94 m	77.56 m

FIGURE 4.7 Cross section of bridge deck. (Architectural drawings provided by Foster + Partners. Image courtesy of Foster + Partners.)

pylons are set into the deck in both the longitudinal and transverse directions to ensure continuity between the metal sheets of the central box girder and those of the walls of the pylons legs and also to provide rigidity by a frame that covers the bearings found on each pier shaft.

4.6.2 Material Used in the Abaqus® Model

In the analysis, C60 is used in the simulation. The metal pylons and deck are made of steels of grades S355 and S460.

4.6.3 3D Modelling Setup

The first step was to set up an AutoCAD 3D wireframe model and import it to ETABS as a DXF file. The model was set up based on the architectural drawings provided by Foster + Partners. It replicates the original structure with reasonable simplification. The 3D model after importation into ETABS is shown in Figure 4.8. In ETABS, the properties and size for all structural members are defined and then converted into Abaqus® INP files using the program of Fu (2009). The 3D Abaqus® model is shown in Figure 4.9.

In the model, the deck of the bridge was modelled with a thin shell member with a thickness of 250 mm, an approximated dimension. Viaduct and individual track members were not modelled. However, the correspondent weights were worked out and added to the model. The truss underneath the deck was modelled using the beam element in Abaqus® (Figure 4.10).

In the model, the piers of the bridge were modelled with four box sections, together with the shell elements, to form the shape of the piers (Figure 4.11).

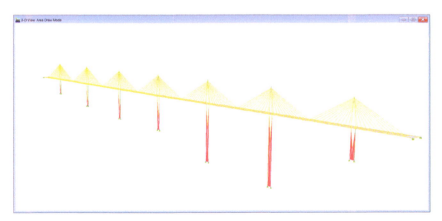

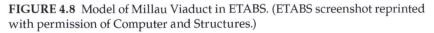

FIGURE 4.8 Model of Millau Viaduct in ETABS. (ETABS screenshot reprinted with permission of Computer and Structures.)

FIGURE 4.9 Contour of the speed after one cable removal. (Abaqus® screenshot reprinted with permission from Dassault Systèmes.)

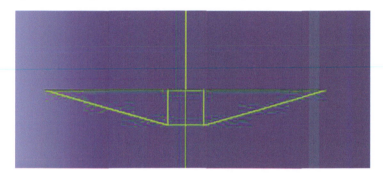

FIGURE 4.10 Cross section of a bridge deck in an Abaqus® model. (Abaqus® screenshot reprinted with permission from Dassault Systèmes.)

FIGURE 4.11 Extruded view of the piers. (Abaqus® screenshot reprinted with permission from Dassault Systèmes.)

4.6.4 *Define the Prestressed Force*

For a cable-stayed bridge, the cables are not slackened, as they are prestressed through connecting to the end of the deck. Therefore, in modelling, it is essential to define the prestressed force at the beginning of the analysis.

In reality, cables are designed only to resist tensile force but not compression force. Therefore, in the Abaqus® input file, the following option is used:

```
*NO COMPRESSION
```

In real construction projects, all of the cables are prestressed. To simulate this prestress force in the Abaqus® input file, the following option is used:

```
*Initial Conditions, type = STRESS
```

In construction practice, the level of prestress in the cable is roughly equal to 60% of its yield stress; therefore, the value 213,000,000 N/mm² was taken in the model.

4.6.5 *Nonlinear Geometric Analysis of the Bridge*

Due to the geometric nonlinearity of the cable, the overall load–displacement relationship of a cable-stayed bridge is nonlinear. Therefore, this nonlinearity needs to be included in the analysis, and nonlinear geometric analysis is required in Abaqus®. The parameter nlgeom = yes in the STEPS command can be used to perform the nonlinear geometric analysis in Abaqus®; therefore, in the input file, the following options can be used:

```
*STEP,INC = 5000,nlgeom = yes,unsymm = yes (the
parameter
*STATIC
0.02
```

The analysis is divided into three steps:

1. **Nonlinear geometric static analysis.** In this step, the prestress forces are redistributed under an external gravity load, such as dead load and live load. The analysis determines the initial geometrical equilibrium under the prestressed force.
2. **Cable removal analysis.** A cable, as highlighted in Figure 4.12, is removed from the model using the same command as shown in Chapter 2.

3. **Dynamic response recording procedure.** The response of the structure after the cable removal is recorded using the command as shown in Chapter 2.

4.6.6 Cable Removal and Load Combination

The time duration of sudden loss of a single cable is required by PTI (2001); therefore, in this study, one cable (ID 35, as highlighted in Figure 4.12) is removed with the same time duration. In accordance with GSA (2003), the following load combination is chosen in the analysis.

$1.0DL + 0.25LL$

where DL is the dead load and LL is the live load.

4.6.7 Major Abaqus® Commands Used in the Simulation

The major commands are shown as follows. Only major steps are demonstrated here. Refer to Chapter 2 for further information.

1. Define the node coordinates (determine the coordinates of the building).

```
*node,nset = Nodes (defining a node set called
  nodes)
1,401.4637,55,0
2,401.4637,55,-94
3,418.4637,55,-94
. . . . . . . .
```

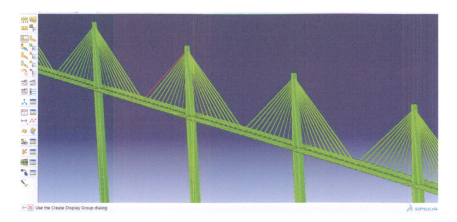

FIGURE 4.12 Cable (ID 35, highlighted) is removed from the model. (Abaqus® screenshot reprinted with permission from Dassault Systèmes.)

2. Define the shell elements (define the slabs and shell of the pier).

```
*element,type = s4r,elset = DECK1
100015,35,672,667,1294 . . . . . . . .
*element,type = s4r,elset = DECK1
100052,35812,36077,36078,35813
. . . . . . . .
```

3. Define the boundary condition (choose all the nodes at the bottom and name them bottomnode).

```
*nset,nset = bottomnode
11
10
. . . . . . .
*boundary (defining the boundary condition).
bottomnode,1,6
```

4. Define the frame element (define the cable and box section for the pier and beams in the deck).

```
*element,type = b31,elset = BOX1 (Box section is
   defined here)
1,515,516
*beam section,section = Box,
   elset = BOX1,material = steel
5,5,2,.3,2,.3
0,-.999080822956353,0 . . . . . .
*element,type = b21h,elset = CABLE199 (cable
   section is defined here)
199,116,532
*beam section,section = CIRC,
   elset = CABLE199,material = Tensiononly (a
   material call Tensiononly is used here, this
   material will be defined in the later part)
.75
0,-.680096835497115,0
```

5. Release definition (define the moment release for the beams).

```
*elset,elset = momrels1
35
. . . . . . .
*elset,elset = momrels2
36
. . . . . . .
*release
```

```
momrels1,s1,M1-M2
momrels2,s2,M1-M2
```

6. Prestress definition (define the prestress for all the cables).

```
*Initial Conditions, type = STRESS
CABLE100 ,   213000000 (as it is explained 60%
  yield stress is applied)
CABLE104 ,   213000000
```

7. Shell section definition (define the shell element for the deck).

```
*shell section,elset = DECK1,material = Concrete
0.25,9
*rebar layer
a252x,50.26e-6,0.200,0.03,s460,,1
a252y,50.26e-6,0.200,0.03,s460,,2
```

8. Define the element set for column removal (define an element set named removal; choose the element with ID 35).

```
*elset,elset = removal
35
```

9. Define the materials for the steel and concrete.
   a. Steel material definition (this sets up the material property for the steel member; only the method for defining no compression is demonstrated; for other steel materials, refer to Chapter 2).

```
*MATERIAL,name = TENSIONONLY
*elastic,type = iso
2.10E+11,0.3,20
*NO COMPRESSION
.... (remaining material definition is same to
  Chapter 2)
```

   b. Concrete material definition (refer to Chapter 2).
10. Define the analysis steps. As discussed in Section 4.6.5, three analysis steps are defined here:
   a. Geometric nonlinear analysis

```
*STEP, INC = 5000,nlgeom = yes, unsymm = yes
*STATIC
0.02
*controls, analysis = discontinuous, field =
  displacement
*controls, parameters = field, field =
  displacement
```

```
0.01,1.0
*controls, parameters = field, field = rotation
0.02,1.0
. . . . . . . .
```

The remaining command is similar to the example in Chapter 2, so refer to it for further information.

b. Cable removal (refer to Chapter 2)
c. Dynamic procedure (refer to Chapter 2)

4.6.8 Simulation Result Interpretation

The model is analysed and the results represented in terms of contour plots and data plots, which are demonstrated here.

4.6.8.1 Contour Plots

After cable 35 was forcibly deleted from the model, the contour plots for variables such as acceleration, velocity, and vertical displacement could be checked. They are shown in Figures 4.13 through 4.15.

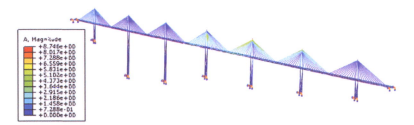

FIGURE 4.13 Contour of acceleration after cable removal. (Abaqus® screenshot reprinted with permission from Dassault Systèmes.)

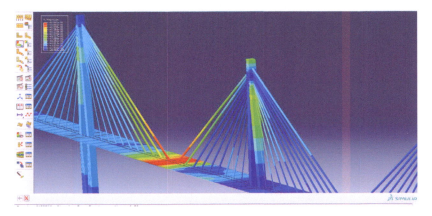

FIGURE 4.14 Spatial velocity contour after cable removal. (Abaqus® screenshot reprinted with permission from Dassault Systèmes.)

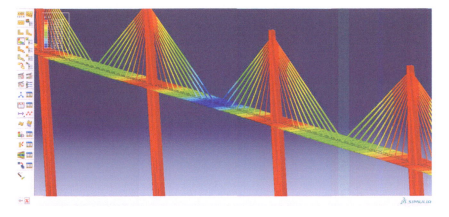

FIGURE 4.15 Contour of vertical displacement after cable removal. (Abaqus® screenshot reprinted with permission from Dassault Systèmes.)

4.6.8.2 *Time History of Certain Parameters* Because cable 35 was forcibly deleted from the model, the axial force of cable 63, which is adjacent to cable 35 (as shown in Figure 4.16), is selected for investigation. The result is shown in Figure 4.17.

Similarly, the axial force of cable 33 (Figure 4.18) is selected for investigation. The result is shown in Figure 4.19.

Similarly, the vertical deflection of the deck near cable 33 can be checked, as shown in Figure 4.20.

4.6.9 *Progressive Collapse Potential Check*

From Figures 4.17 and 4.20, it can be seen that at 1 second, the cable was removed; however, the axial force of cables 33 and 63 did not have

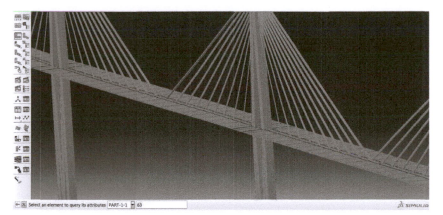

FIGURE 4.16 Cable 63 is selected (highlighted). (Abaqus® screenshot reprinted with permission from Dassault Systèmes.)

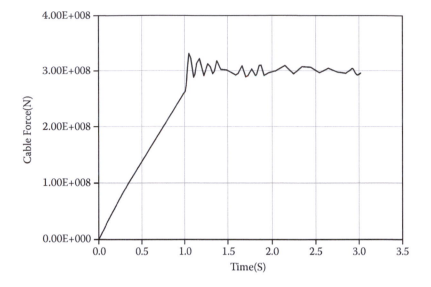

FIGURE 4.17 Cable force of cable (ID 63).

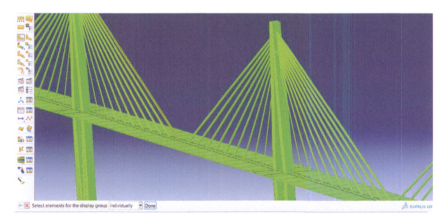

FIGURE 4.18 Cable 33 is selected (highlighted). (Abaqus® screenshot reprinted with permission from Dassault Systèmes.)

a dramatic increase, and the two cables did not fail. The vertical deflection increased from 0.16 minute at 1 second to 0.28 minute after the cable removal. Therefore, the progressive collapse potential is small.

4.6.10 Five-Cable Removal Check

As required by PTI (2001), in addition to the loss of one cable, designers should also check the removal of multiple cables. Therefore, in this exercise, five cables (their locations are shown in Figure 4.21) were removed simultaneously.

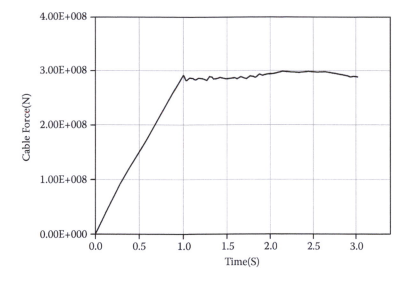

FIGURE 4.19 Cable force of cable (ID 33).

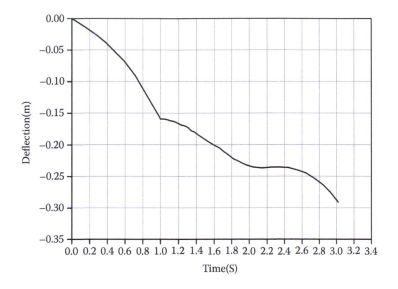

FIGURE 4.20 Vertical deflection of the deck near the removed cable.

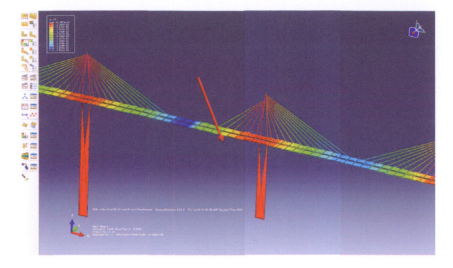

FIGURE 4.21 Contour plot of vertical deflection after the removal of five cables (the locations of the five cables are indicated). (Abaqus® screenshot reprinted with permission from Dassault Systèmes.)

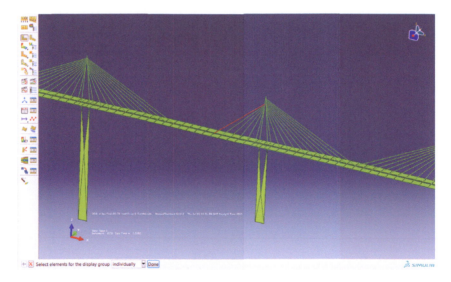

FIGURE 4.22 Cable 175 is selected. (Abaqus® screenshot reprinted with permission from Dassault Systèmes.)

After the analysis, cable 175, which is adjacent to the removed cables, as shown in Figure 4.22, is selected for investigation. From Figure 4.23, it can be seen that the cable force dramatically increases to 45,000 kN; therefore, rupture of this cable will occur and progressive collapse is likely to be triggered.

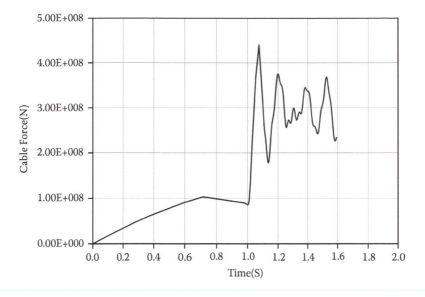

FIGURE 4.23 Cable force of cable (ID 175).

References

AASHTO (American Association of State Highway and Transportation Officials). 2007. Standard specifications for highway bridges. Washington, DC: AASHTO.

Billah, K., and Scanlan, R. 1991. Resonance, Tacoma Narrows Bridge failure, and undergraduate physics textbooks. *American Journal of Physics*, 59(2), 118–124.

European Committee for Standardization. 2003. Eurocode 1: Actions on structures. Part 2: Traffic loads on bridges. EN 1991-2. European Committee for Standardization.

FIB (International Federation for Structural Concrete). 2005. Acceptance of stay cable systems using prestressing steels. Lausanne: FIB.

Fu, F. 2009. Progressive collapse analysis of high-rise building with 3-D finite element modelling method. *Journal of Constructional Steel Research*, 65, 1269–1278.

GSA (General Services Administration). 2003. Progressive collapse analysis and design guidelines for new federal office buildings and major modernization projects. Washington, DC: GSA.

Kawashima, K., Takahashi, Y., Ge, H., Wu, Z., and Zhang, J. 2009. Reconnaisance report on damage of bridges in 2008 Wenchuan, China, earthquake. *Journal of Earthquake Engineering*, 13(7).

Lynch, D. 2011. Hanger work suspected in Indonesia bridge collapse. *New Civil Engineer*, December 1.

Lynch, D. 2012. Basic engineering errors led to Indonesia bridge collapse. *New Civil Engineer*, January 26.

PTI (Post-Tensioning Institute). 2007. *Recommendations for Stay Cable Design, Testing and Installation.* 5th ed. Phoenix, AZ: Cable-Stayed Bridge Committee.

JRC-Ispra, Seminar on 'Bridge Design with Eurocodes'. 2012. Organized and supported by European Commission, Russian Federation, European Committee for Standardization.

Starossek, U. 2006. Progressive collapse of bridges—Aspects of analysis and design. Invited lecture at International Symposium on Sea-Crossing Long-Span Bridges, Mokpo, South Korea, February 15–17.

Zoli, T.P., and Steinhouse, J. Some considerations in the design of long span bridges against progressive collapse. https://www.pwri.go.jp/eng/ujnr/tc/g/pdf/23/23-2-3zoli.pdf.

CHAPTER 5

Fire-Induced Building Collapse

5.1 Introduction

For structures, especially steel structures, fire incidents are one of the major causes of damage and even collapse of buildings. It has been estimated that from 2012 to 2013, there were about 64,000 fire outbreaks in buildings around the world. The collapse of World Trade Center 7 (WTC7), which was caused by fire set by falling debris from the World Trade Center, makes fire-induced building collapse a pressing issue that design engineers need to tackle. It also brought the attention of researchers to the study of the structural behaviour of buildings under fire conditions and effective design methods to prevent or delay the collapse of buildings.

In this chapter, basic fire design knowledge and the collapse mechanisms of building under fire are introduced. The methods for designing a building to prevent progressive collapse are discussed. At the end of the chapter, a modelling example of WTC7 under fire is demonstrated using the general-purpose program Abaqus®.

5.2 Basic Knowledge of Fire

Before we discuss the behaviour of structures under fire and possible methods to delay or prevent their collapse, it is worth reviewing some basic fire knowledge for a better understanding of the behaviour of a building under fire.

5.2.1 *Fire Development Process and Fire Temperature Curve*

As shown in Figure 5.1, the development of fire can be divided into five stages: growth phase, flashover phase, fully developed phase, decay phase, and final extinction. Flashover is the transition from localized fire to a fire that consumes the whole room. However, if a room has very large window openings, too much heat may flow out

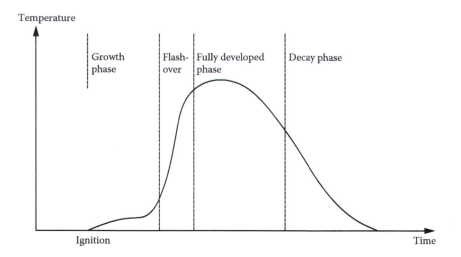

FIGURE 5.1 Temperature–time curve for full process of fire development.

the windows for flashover to occur. After flashover, the fire is often referred to as a postflashover fire. The rate of combustion depends on the size and shape of the ventilation openings.

To represent the above fire development process in structural fire design, fire temperature curves for the atmosphere are used. There are two main fire temperature curves that engineers can use directly in their structural fire analysis: the standard fire temperature–time curve and the parametric temperature–time curve.

5.2.1.1 Standard Fire Temperature–Time Curve The standard fire temperature curve is defined as shown in Figure 5.2. It can be seen that in this temperature curve, the cooling phase is not included. The curve defines the heating condition for fire tests on structural members; however, in some research, it is still used for fire analysis, as it is convenient and more conservative. BS 476: Part 20 (BSI, 1987) gives the formula to calculate the temperature:

$$T = 345 \log 10 \,(8t + 1) + 20 \tag{5.1}$$

where T is the mean furnace temperature (in °C) and t is the time (in min), up to a maximum of 360 min.

EN 1991-1-2: Eurocode 1, Part 1.2 (BSI, 2002a), also gives a similar formula to work out the standard temperature–time curve.

5.2.1.2 Parametric Temperature–Time Curve The parametric fire is defined in Annex A of EN 1991-1-2: Eurocode 1, Part 1-2 (BSI,

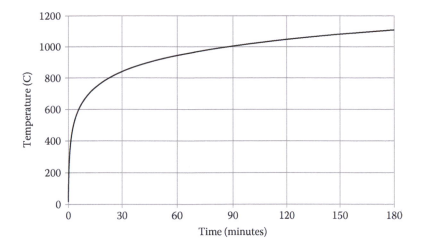

FIGURE 5.2 Standard fire temperature–time curve. (From BSI, Incorporating amendment no. 1, Fire tests on building materials and structures, Part 20: Method for determination of the fire resistance of elements of construction (general principles), BS 476-20, London: BSI, 1987, Figure 2, p. 33.) Permission to reproduce extracts from British Standards is granted by BSI. British Standards can be obtained in PDF or hard-copy formats from the BSI online shop (www.bsigroup.com/Shop) or by contacting BSI customer service for hard copies only (telephone: +44 (0)20 8996 9001, email: cservices@bsigroup.com).

2002a), as shown in Figure 5.3. When determining worst-case conditions for the compartment, the main variable is percentage of openings.

The parametric temperature–time curve in the heating phase is given by EN 1991-1-2: Eurocode 1, Part 1-2 (BSI, 2002a):

$$\Theta_g = 20 + 1325(1 - 0.324e^{-0.2t^*} - 0.201e^{-1.7t^*} - 0.472e^{-19t^*}) \quad (5.2)$$

where Θ_g is the gas temperature in the fire compartment, $t^* = \Gamma t$ with t time and

$$\Gamma = [O/b]^2/[0.04/1160]^2$$

where O is the opening factor,

$$O = A_v \cdot H_w^{\wedge 0.5}/A_t$$

where A_t is the total internal surface area of the compartment (m²), A_v is the area of ventilation (m²), H_w is the height of openings (m), and b is the thermal diffusivity, $100b[2000(J/m^2\,s^{1/2}\,K)]$.

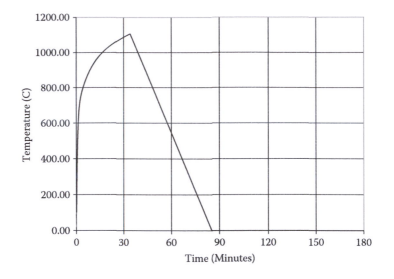

FIGURE 5.3 Parametric temperature–time curve. (From BSI, Eurocode 1: Actions on Structures, Part 1-2: General actions—Actions on structures exposed to fire, BS EN 1991-1-2, London: BSI, 2002.) Permission to reproduce extracts from British Standards is granted by BSI. British Standards can be obtained in PDF or hard-copy formats from the BSI online shop (www.bsigroup.com/Shop) or by contacting BSI customer service for hard copies only (telephone: +44 (0)20 8996 9001, email: cservices@bsigroup.com).

The maximum temperature, Θ_{max}, in the heating phase happens for $t^* = t^*_{max}$,

$$t^*_{max} = t_{max} \cdot \Gamma$$

with $t_{max} = (0.2 \cdot 10^{-3} \cdot q_{t,d}/O)$, or t_{lim}.

It can be seen that the parametric temperature–time curve is closer to real fire development, as shown in Figure 5.1. Therefore, it is used in most fire analysis.

5.2.2 Heat Transfer and Thermal Response of Structural Members

Heat transfer consists of three processes: conduction, convection, and radiation. Conduction is the mechanism of heat transfer in solid materials, in steady-state situations. Convection is heat transfer by the movement of fluids, either gases or liquids. Radiation is the transfer of energy by electromagnetic waves. The transfer of the heat through the above process can be worked out based on a formula provided by Eurocode 3 (BSI, 2005a).

Through the heat transfer process, heat can be transferred to the structural members. As long as the atmosphere fire temperature (also called gas temperature) (determined using either the standard temperature curve or the parametric temperature curve) is known, the thermal response of each structural member can be worked out. The thermal response of structural elements can be obtained from fire tests such as those given in BS 476: Parts 20–22 (BSI, 1987) and ISO 834-1 (ISO, 1999), which utilise a standardized temperature–time curve. In addition, BS EN 1993-1-2: Eurocode 3 (BSI, 2005a) and BS EN 1994-1-2: Eurocode 4 (BSI, 2005b) give the formula to work out an increase of temperature for both internal unprotected and protected steelwork and concrete materials. These formulas are based on the principles of heat transfer. Designers can use them to work out the thermal response of structural steel members.

For Unprotected steel Section, the increase of temperature in in a small time interval is given by BS EN 1993-1-2: Eurocode 3 (BSI, 2005a) and BS EN 1994-1-2: Eurocode 4 (BSI, 2005b) as follows:

$$\Delta\theta_{a,t} = k_{sh}\frac{A_m/V}{c_a\rho_a}h_{net}\Delta t \tag{5.3}$$

where,

$\Delta\theta_{a,t}$ is increase of temperature

A_m/V is the section factor for unprotected steel member

c_a is the specific heat of steel

ρ_a is density of the steel

h_{net} is the designed value of the net heat flux per unit area

Δt is the time interval

k_{sh} is the correction factor for the shadow effect

For protected steel Section, the increase of temperature in in a small time interval is given by BS EN 1993-1-2: Eurocode 3 (BSI, 2005a) and BS EN 1994-1-2: Eurocode 4 (BSI, 2005b) as follows:

$$\Delta\theta_{a,t} = \left\{\frac{\lambda_p/d_p}{c_a\rho_a}\frac{A_p}{V}\left(\frac{1}{1+\Phi/3}\right)(\theta_{g,t}-\theta_{a,t})\Delta t\right\}-\left\{\exp(\Phi/10)-1\right\}\Delta\theta_{g,t} \tag{5.4}$$

where,

$$\Phi = \frac{c_p\rho_p}{c_a\rho_a}d_pA_p/V$$

$\theta_{a,t}$ is temperature of the steel at time t

$\Delta\theta_{a,t}$ is increase of temperature

$\Delta\theta_{g,t}$ is the ambient gas temperature at time t

$\Delta\theta_{g,t}$ is increase of the ambient gas temperature

A_p/V is the section factor for protected steel member

c_a is the specific heat of steel

c_p is the specific heat of fire protection material

ρ_a is density of the steel

ρ_p is density of the fire protection material

d_p is thickness of the fire protection material

λ_p is the thermal conductivity of the fire protection material

Δt is the time interval

5.2.3 *Material Behaviour at Elevated Temperatures*

The material properties of steel and concrete start to lose strength at elevated temperatures. Figure 5.4 shows the stress–strain curves of steel at different temperatures from EN 1994-1-2: Eurocode 4 (BSI, 2005b). The loss of strength can be illustrated by the amount of stress

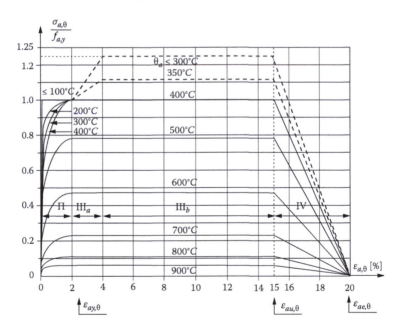

FIGURE 5.4 Graphical presentation of the stress–strain relationships of structural steel at elevated temperatures; strain hardening is included. (From BSI, Eurocode 4: Design of composite steel and concrete structures, Part 1-2: General rules, Structural fire design, BS EN 1994-1-2, London: BSI, 2005, Figure A.1.) Permission to reproduce extracts from British Standards is granted by BSI. British Standards can be obtained in PDF or hard-copy formats from the BSI online shop (www.bsigroup.com/Shop) or by contacting BSI customer service for hard copies only (telephone: +44 (0)20 8996 9001, email: cservices@bsigroup.com).

that the member is able to withstand before reaching the 2% strain. There is a significant drop of strength between 400°C and 700°C. It can be seen that when steel is heated up to 800°C, it is only at 11% of its initial strength.

Reduction factors are used to account for the degradation in yield strength, elastic modulus, and proportional limit. They are defined as proportions of values at elevated temperatures to those at ambient temperature. Figure 5.5 shows the reduction factor for structural steel members defined by EN1994-1-2: Eurocode 4 (BSI, 2005b).

Figure 5.6 shows the stress–strain relationship of concrete from EN 1994-1-2: Eurocode 4 (BSI, 2005b). It can be seen that the stress–strain relationships of concrete exhibit a linear response, followed by a parabolic response until peak stress, and then a steep, descending slope prior to failure. The temperature shows a significant effect on the stress–strain relationships of concrete.

In structural fire analysis, engineers should use the above material properties in their analysis, which are demonstrated in Section 5.7.

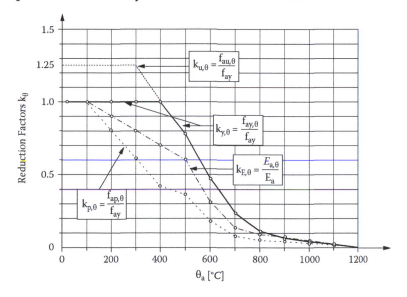

FIGURE 5.5 Reduction factors for stress–strain relationships allowing for strain hardening of structural steel at elevated temperatures. (From BSI, Eurocode 4: Design of composite steel and concrete structures, Part 1-2: General rules, Structural fire design, Incorporating corrigendum July 2008, BS EN 1994-1-2, London: BSI, 2008, Figure A.2.) Permission to reproduce extracts from British Standards is granted by BSI. British Standards can be obtained in PDF or hard-copy formats from the BSI online shop (www.bsigroup.com/Shop) or by contacting BSI customer service for hard copies only (telephone: +44 (0)20 8996 9001, email: cservices@bsigroup.com).

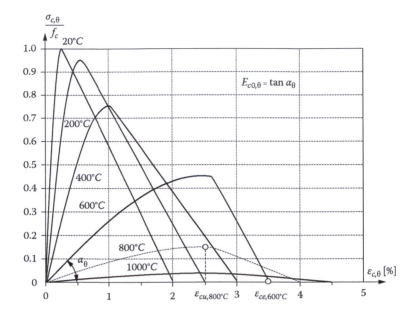

FIGURE 5.6 Compressive concrete material behaviour. (From BSI, Eurocode 4: Design of composite steel and concrete structures, Part 1-2: General rules, Structural fire design, Incorporating corrigendum July 2008, BS EN 1994-1-2, London: BSI, 2008, Figure B.1.) Permission to reproduce extracts from British Standards is granted by BSI. British Standards can be obtained in PDF or hard-copy formats from the BSI online shop (www.bsigroup.com/Shop) or by contacting BSI customer service for hard copies only (telephone: +44 (0)20 8996 9001, email: cservices@bsigroup.com).

5.2.4 Fire Protection Method

There are two main categories of fire protection methods: active control system and passive control system. For active control, the protections are based on the action taken by a person or an automatic device, such as a sprinkler. For a passive control, the fire protection systems are built into the structure of the building, such as intumescent paint, spray, or board protection of the structural steel members.

5.3 Fire Incidents around the World

In this section, several fire incidents that have occurred around the world are introduced. In some of the incidents, the collapse of the whole building or a partial collapse of the building was triggered. However, in other incidents, no collapse was observed. The reasons for the collapse and the collapse mechanisms for buildings under fire are explained in detail in Section 5.4.

5.3.1 WTC7 Collapse (Progressive Collapse Is Triggered)

As shown in Figure 5.7, World Trade Center 7 was a 47-storey commercial building located close to the Twin Towers. It had an irregular trapezoid shape (Figure 5.8) and was a steel composite frame building. The main lateral stability and the gravity resistance system used the moment connection frames with 21 internal columns to form a rectangular building core, another lateral stability system.

As mentioned earlier, the collapse of WTC7 was mainly due to the fire ignited as the result of the debris from the collapse of WTC1, rather than the aircraft collision. There were both passive and active fire protection systems in WTC7 (NIST NCSTAR, 2008); the passive fire protections used sprayed fire-resistant material (SFRM) on the structural steel and metal decking for the floors. The active fire protection system consisted of sprinklers inside the building.

It was found from the NIST NCSTAR (2008) investigation that the sprinklers were not functional due to the cutoff of the main water

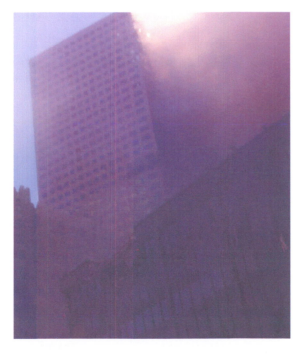

FIGURE 5.7 WTC7 collapse. (From http://upload.wikimedia.org/wikipedia/commons/thumb/0/0e/Wtc7onfire.jpg/511px-Wtc7onfire.jpg. *Courtesy of the Prints and Photographs Division, Library of Congress, public domain.*)

supply during the accident; therefore, fire became the major reasons for the collapse of the building.

Figure 5.8 shows the typical floor layout of WTC7. According to NIST NCSTAR (2008), the collapse was triggered by the buckling of an interior column in the northeast region of the building, as shown in Figure 5.9, which led to a floor failure and buckling of adjacent internal columns progressively. This further resulted in the buckling

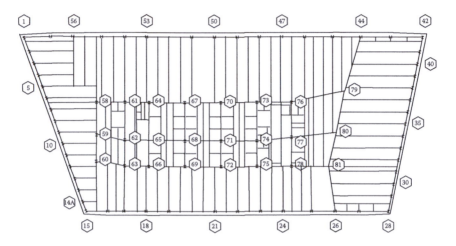

FIGURE 5.8 Typical floor of WTC7. (Permission to reproduce and derive from NIST NCSTAR, Federal building and fire safety investigation of the World Trade Center disaster, Final report on the collapse of World Trade Center Building 7, Gaithersburg, MD: NIST, U.S. Department of Commerce, 2008, Figure 1-5.)

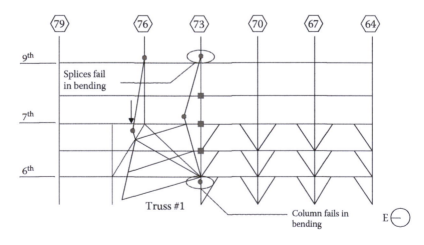

FIGURE 5.9 Collapse mechanism of WTC7. (Permission to reproduce from http://www.wtc7.net/nistreport.html on 28/07/2015.)

of the column in the horizontal progression and the buckling of exterior columns as the failed building core moved downward.

Global collapse then started when the entire building above the buckled region moved downward like a pancake. The collapse is shown in Figure 5.7.

5.3.2 *Windsor Tower (Partial Collapse Is Triggered)*

As shown in Figure 5.10, the Windsor Tower is 106 metres high with 32 floors. In 2005, a fire started on the 21st floor and quickly spread throughout the entire building, leading to extensive progressive collapse above the 17th floor.

FIGURE 5.10 Fire in Windsor Tower. (This file is licensed under the Creative Commons Attribution—Share Alike 3.0 free licence, http://commons.wikimedia .org/wiki/File:TorreWindsor1.JPG.)

The tower is built with a reinforced concrete (RC) core for lateral stability and waffle slabs supported by internal RC columns and steel beams; the perimeter is supported by steel columns. The perimeter columns and internal steel beams were left unprotected above the 17th floor level in accordance with the Spanish building code at the time of construction. By the time the fire broke out, the fire protection for all steelwork below the 17th floor had been completed except a proportion of the 9th and 15th floors.

The studies from Dave (2005) and NILIM (2005) show that the fire resulted in the simultaneous buckling of the unprotected steel perimeter columns of several floors, triggering the collapse of the floor slabs above the 17th floor.

For the floors below the 17th floor, except for the unprotected columns at the 9th and 15th floors, which also buckled, no structural collapse was observed. The investigation of University of Manchester (2005) found that the reinforced concrete central core, columns, waffle slabs, and transfer structures performed very well in such a severe fire. The structural integrity and redundancy of the remaining parts of the building provided the overall stability of the building.

5.3.3 Beijing Television Cultural Centre Fire (Partial Collapse)

As shown in Figure 5.11, in 2009, the entire building of the Beijing Television Cultural Centre caught fire; it was put out 6 h later. The

FIGURE 5.11 CCTV building on fire. (From Wikipedia, free licence, http://commons.wikimedia.org/wiki/File:CCTV_new_headquarters_Fire_20090209.jpg.)

cause of the fire was the illegal use of highly explosive fireworks at the construction site. The fire started after a shell from the fireworks landed on the roof of the uncompleted construction (Jacobs, 2009). Fortunately, only local collapse was observed; the main building did not collapse.

5.3.4 Cardington Fire Test

The Cardington fire test was the first full-scale fire test in history. Therefore, it is worth introducing here. It was conducted by British Steel and Building Research Establishment (BRE, 1999). Six fire tests on an eight-storey typical braced steel office building were performed at Cardington in the UK.

Test 1 was a restrained beam test. An unprotected 9 m long internal beam and supported slab were heated by a gas-fired furnace in the middle until the temperature reached 800°C to 900°C through the section profile. Connections were still at ambient temperature. Yielding and local buckling at both ends of the test beam were also observed during the experiment. The lower flanges at the ends of the beam were distorted as restraining forces occurred due to thermal expansion against the web of the column section.

Test 2 was a plane frame test. It was designed to investigate the primary beams and columns along gridline B, which supported the fourth floor. The primary and secondary beams and top 800 mm of the columns were left unprotected. It was observed that the exposed parts of the columns were squashed at approximately 670°C. This may lead to the floors above the fire compartment turning unstable. Therefore, it was suggested by BRE (1999) that the columns be fully protected along the entire length to limit damage to the fire compartment area only.

Test 3 was a corner compartment test to investigate the behaviour of the composite floor under fire, especially the membrane effect. All structural members were left unprotected apart from columns, column-to-beam connections, and external perimeter beams. The maximum recorded steel temperature was 935°C. Extensive buckling was noticed at the beam-to-column connections. The end of an internal secondary beam, which was connected to a primary beam, buckled locally due to axial restraint from adjacent members. However, no local buckling occurred at the other end of the beam, which was connected to an external beam. This was because the thermal expansion of the secondary beam caused the external beam to twist, resulting in insufficient restraint to cause local buckling.

Test 4 was another corner compartment test. Only columns were protected. Windows and doors were closed, leading to the development of the fire being restricted by a low level of oxygen. Then the temperature of the fire dropped after the initial rise, and the fire continued. Flashover did not occur until two windows were destroyed. It was recorded that the maximum steel temperature was 903°C. The compartment wall was found to affect how unprotected beams perform. When the wall was removed, distortional buckling occurred over most of the beam length. This was due to the positioning of the wall, causing a high thermal gradient through the section profile of the beam.

Test 5 was a large (340 m²) compartment test that was between the second and third floors. A fire resistance wall was constructed along the full width of the building. Unlike test 4, enough ventilation was allowed for the fire to develop. All of the steel beams were left unprotected. The maximum atmosphere temperature and steel temperature recorded were 746°C and 691°C, respectively. The fire was not very severe because it lasted longer with lower temperatures. Many beam-to-beam connections were found to have locally buckled and many endplate connections fractured down one side after cooling.

Test 6 was a simulated office test, where a more realistic open-plan office fire scenario was simulated using office furniture as the consuming fuel in a compartment area of 135 m². Only columns and beam-to-column connections were protected. The maximum steel temperature was 1150°C. No signs of failure were observed, but there was extensive cracking during the latter phase of cooling.

5.4 Collapse Mechanisms of Buildings in Fire

From the above incidents and tests, it can be seen that the structural system and effective fire protection regimes are key to preventing the collapse of buildings in fire. Therefore, it is worthwhile to further discuss the collapse mechanisms of buildings in fire in this section.

5.4.1 Floor System Slab Failure and Membrane Effect

When using yield line analysis to estimate the upper-bound bending resistance of slabs in the Cardington tests, it was found that the applied load on the Cardington test (BRE, 1999) structure was much higher than the upper bound solution. This phenomenon can be explained by so-called membrane action. Since the deflection of the

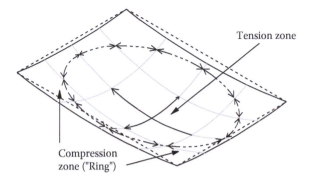

FIGURE 5.12 Tensile membrane action of slabs.

slab was large due to fire, membrane action occurred and contributed to resisting the applied load. Because the weakening beams and large deflections resulted in a change in the load transfer mechanism, the slabs were able to bridge over the fire-damaged supporting beams and transfer load to the undamaged parts of the structure, through membrane action. As we can see from Figure 5.12, when a slab is in tensile membrane action, the tensile force in the reinforcement is resisted by a compressive ring formed within the slab, around its edges, provided that the edges are vertically supported. The tensile membrane action will only be dominant under large deflections, distributing the loads to the adjacent members and providing a much higher load-carrying capacity than the normal bending capacity of the slab.

This membrane action can also be demonstrated in the three-dimensional (3D) finite-element model. Fu (2015) has conducted research on the whole-building response of a steel composite tall building in fire. In his research, a 20-storey building was simulated using Abaqus®, and the fire temperature was applied to levels 9–11. The model is shown Figure 5.13; for detailed modelling techniques, refer to Fu (2015) and Section 5.7 of this chapter. Figure 5.14 shows the contour of the vertical deflection. If we check the axial force inside slab A3-A4-B4-B3 at floor 10, it can be seen that in the elements at the centre of the slab, the axial force is in tension, as shown in Figure 5.15. However, in Figure 5.16, which shows the axial force at the edge element of slab A3-A4-B4-B3, it is found that it is mainly in compression.

Figure 5.17 is the plastic strain of the concrete slabs, which also indicates the location of the crack formed in the concrete. We notice that the large plastic strain is observed at the edge slabs, which indicates the crack pattern of the concrete slabs.

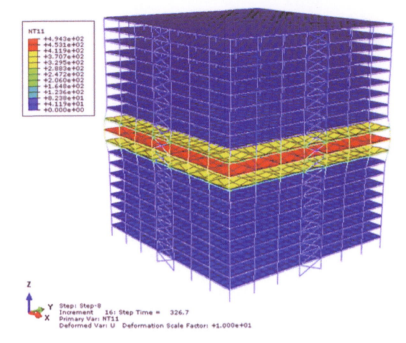

FIGURE 5.13 3D Abaqus® model of tall building with fire set on three storeys. (Abaqus® screenshot reprinted with permission from Dassault Systèmes.)

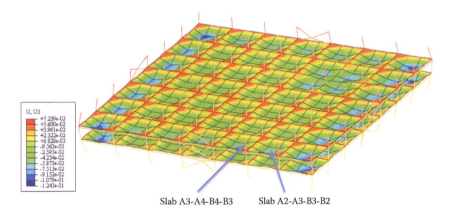

Slab A3-A4-B4-B3   Slab A2-A3-B3-B2

FIGURE 5.14 Contour of vertical deformation of slabs on floors 9 and 10. (Abaqus® screenshot reprinted with permission from Dassault Systèmes.)

Bailey and Moore (2000) developed a method using a simple energy approach to calculate the load-carrying capacity of a composite flooring system. A failure criterion that considered the mechanical strains and thermal effects of the system was proposed. Bailey's method can predict the real performance of a slab in fire by

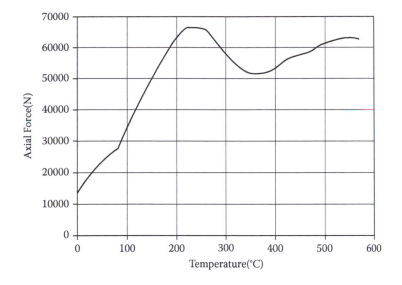

FIGURE 5.15 Axial force at the centre of slab A3-A4-B4-B3 on floor 10.

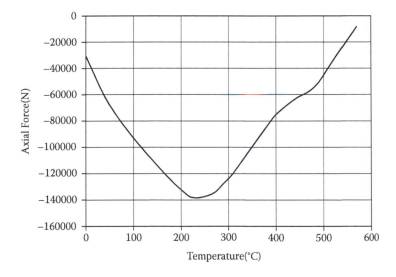

FIGURE 5.16 Axial force at the perimeter of slab A3-A4-B4-B3 on floor 10.

accounting for the membrane action of floor slabs and considering the grillage of the beams and floor slab to act as a unit.

5.4.2 Structural Steel Beam Failure

The response of the beams in structures subject to fire is strongly influenced by the restraint provided by the adjacent structural members. In the standard fire test, the excessive deflection experienced

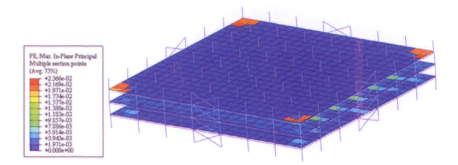

FIGURE 5.17 Plastic strain of the concrete slab on floors 9 to 11. (Abaqus® screenshot reprinted with permission from Dassault Systèmes.)

by steel beams is typical. In the Cardington tests (BRE, 1999), even though the temperatures of the bottom flanges of these beams exceeded 800°C, the excessive deflection behaviour did not happen to the unprotected composite beams due to restraint from adjacent members (Wang, 2000). One of the main failure modes of the beam is bending failure. Overall buckling of the beam is not a major issue due to the restraint from the slabs. As indicated in BRE (1999), local buckling, particularly in the negative bending moment region, is significant, as this will affect the behaviour of the columns.

5.4.3 *Structural Steel Column Failure*

The failure of columns has been identified as a major reason for the collapse of structures under fire. The key design target is to prevent column failure, as it will result in the collapse of buildings. Research by Talamona et al. (1997) and Wang (2004) shows that the restraints play a vital role in the behaviour of columns. It was found by BRE (1999) that overall buckling, squashing, bending, local buckling, and lateral torsional buckling are major failure mechanisms for columns under fire.

5.4.4 *Structural Steel Connections*

Contrary to what has been traditionally assumed, the Cardington fire tests (BRE, 1999) showed that the connections are more vulnerable in fire. In some existing projects, connections not designed for thermal effects become one of the reasons for the collapse of a building. The connections play a significant role and will influence the behaviour of the whole frames in fire. In the design, the connections need to be designed according to their type, such as flush endplate connection, fin plate connection, and extended endplate connection.

5.5 Structural Fire Design to Prevent Building Collapse

5.5.1 *Current Design Code*

Design codes for fire safety in buildings can be either a prescriptive type or performance-based type. There are major structural fire design codes, such as Eurocode 3, Part 1-2 (BSI, 2005a); Eurocode 4, Part 1-2 (BSI, 2005b); BS 5950, Part 8 (BSI, 1990); and Eurocode 2, Part 1-2 (BSI, 2004). There is also structural fire design guidance such as by the Institution of Structural Engineers (2007).

The main objective for structural fire design is to make sure that in the event of an outbreak of fire, the load-bearing capacity of the building will continue to function until all occupants have escaped or been assisted to escape from the building. The main purpose is to achieve life safety, not collapse prevention. Therefore, no detailed guidance on preventing fire-induced collapse is available so far. However, the lessons from WTC7 show that effective measures to delay or prevent the collapse of structures are imperative.

5.5.2 *Design Recommendation*

NIST NCSTAR (2008) has recommendations to prevent the collapse of tall buildings under fire. They are summarised as follows:

- Explicit evaluation of the fire resistance of structural systems in buildings under worst-case fire designs should be made in case of active fire protection systems becoming ineffective.
- The effects of thermal expansion in long-span floor systems, connections not designed for thermal effects, asymmetric floor framing, and composite slabs should be taken into consideration.
- The performance and redundancy of active fire protection systems should be enhanced to accommodate higher-risk buildings.
- Increased structural integrity: The standards for estimating the load effects of potential hazards (e.g., progressive collapse) and the design of structural systems to mitigate the effects of those hazards should be improved to enhance structural integrity.

5.6 Structural Fire Analysis

There are two categories of natural fire models from BS EN 1991-1-2: Eurocode 1 (BSI, 2002a): simplified fire models and advanced fire models. The techniques introduced here are advanced fire models, as they can model gas properties, mass exchange, and energy exchange.

5.6.1 Zone Model

Zone models are simple computer models that divide the considered fire compartment into separate zones, where the conditions of each zone are assumed to be uniform. The models define the temperature of the gases as a function of time by considering the conservation of mass and energy in the fire compartment. Two-zone models are used for preflashover fires, whereas one-zone models are used for postflashover fires.

5.6.2 CFD Model

Computational fluid dynamics (CFD) models can analyse fluid flow, heat transfer, and associated phenomena by solving the fundamental equations of fluid flow. These equations represent the mathematical statements of the conservation laws of physics. The model can provide information at numerous points within the compartment relating to temperature, velocity, toxic content, and visibility. Therefore, this is one of the most accurate ways to model fire development. However, the skills and knowledge required for using CFD models are very demanding.

5.6.3 Finite-Element Method Using the Fire Temperature Curve

The finite-element method is one of the easiest for engineers to use. Using the fire temperature curve introduced in Section 5.2, the response of buildings under fire can be simulated by including relevant material properties at elevated temperatures. The method is demonstrated in the next section using Abaqus®.

5.7 Modelling Example of Progressive Collapse Analysis of WTC7 under Fire Using Abaqus®

In this section, the modelling method of the whole-building behaviour analysis of tall buildings under fire will be demonstrated. The building is first built in Abaqus® to perform structural fire analysis to identify the failure members, and then using a similar procedure, introduced in Chapter 2, member removal analysis can be performed to determine the progressive collapse potential of the structure. As the purpose of this modelling example is to demonstrate how to model the global behaviour of WTC7 under fire, some reasonable simplifications are made in the analysis.

5.7.1 Prototype Building

The prototype building WTC7 was introduced in Section 5.3.1. The model was first set up in ETABS (CSI, 2008), as shown in Figure 5.18. Using the program designed by Fu (2009), the model was converted to Abaqus® INP files, and the fire analysis was performed in Abaqus®. The layout of the building is shown in Section 5.3.1.

5.7.2 Modelling Procedures

Structural performance in fire is mainly affected by different thermal regimes and the degree of restraint provided by the main components of the structure, such as beams and columns. The 3D model in Abaqus® is shown in Figure 5.19.

5.7.2.1 Element Selection Beam elements in Abaqus® are used to model columns and beams. The disadvantage of beam elements are that they cannot predict local buckling, as it is assumed that the

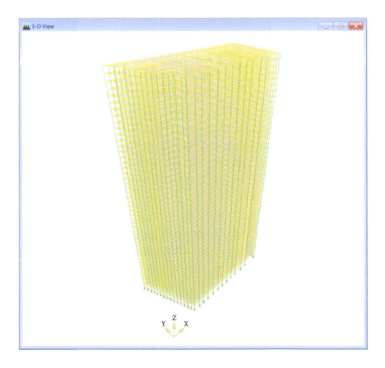

FIGURE 5.18 3D model of tall building built in ETABS. (ETABS screenshot reprinted with permission of CSI.)

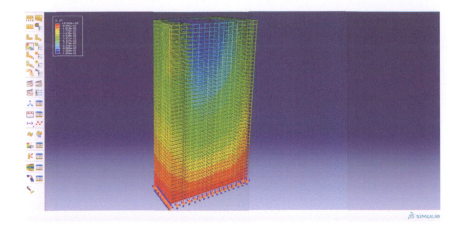

FIGURE 5.19 3D Abaqus® model of WTC7. (Abaqus® screenshot reprinted with permission from Dassault Systèmes.)

beam retains its shape at each cross section. The only way to model local buckling is to model the web and flanges using a shell element; however, this dramatically increases the computational cost. An investigation by BRE (1999) found only a small difference on the deflection between these two modelling techniques. Therefore, to simplify the analysis, local buckling effect is ignored and beam elements are used here. This simplification is reasonable as it is to stimulate the global behavior of the building.

Structural steel beams and columns are represented using a number of beam elements representing the main axial and bending behaviour of each member. Steel beams are assumed to be pin-connected to the steel columns. All beams are modelled on the centreline of the structural member, and the concrete slab is modelled on the centreline of the concrete slab. This ensures the correct degree of restraint to each structural member, particularly on the edge of the building.

The overall buckling of the beams and columns can be included by subdividing each beam or column into several small segments along the actual length. The research by Yang et al. (2010) shows that normally four segments are sufficient to model the global buckling of the beams or columns.

Shell elements are used to model floor slabs. They can be used to model two-dimensional stress states, including both membrane and flexural effects. Integrating through the thickness of the element allows the variation of the properties to be included. Reinforcement inside the slabs was represented as a smeared layer in each shell element using the *REBAR element and was defined in both slab directions. The beam and shell elements were coupled together using rigid beam constraint equations to give the composite action between the beam elements and concrete slabs.

5.7.2.2 Material Constitutive Models under Fire   As mentioned earlier, due to the fire conditions, the stiffness and strength of steel and concrete deteriorate at elevated temperatures. Therefore, an accurate material constitutive model is essential for modelling the behaviour of structures under fire.

5.7.2.2.1 Material Properties of Concrete Slab at Elevated Temperatures Concrete material properties of all slabs were represented using a concrete damaged plasticity model from Abaqus®. Material properties were assumed to vary with temperature using the relationships outlined in EN 1994-1-2 (BSI, 2005b) (Figure 5.6). Reinforcement meshes in the concrete slab were represented using the rebar function of the concrete shell elements.

The reinforcement was assumed to be located 30 mm from the top of the slab with a yield stress of 460 N/mm² and with material properties varying with temperature, as outlined in BS EN 1993-1-2 (BSI, 2005a) and shown in Figure 5.4. The contribution of the structural metal deck is conservatively ignored.

5.7.2.2.2 Material Properties of Structural Steel Members at Elevated Temperatures Steel beam material properties were represented using a Von Mises elastoplastic material model with a yield stress of 355 N/mm². Material properties were assumed to vary with temperature using the relationships outlined in BS EN 1993-1-2 (BSI, 2005a) (Figure 5.4). All of the beams and columns were designed with intumescent protection to resist a 2 h fire.

5.7.2.3 Mesh, Gravity Load, and Boundary Conditions   Generally, the larger the number of finite elements, the more accurate the estimate of the structural response, but the analysis time will increase. A balance needs to be made between the number of elements and the required accuracy. Normally, a sensitivity analysis of the mesh selection can be made. As the loading condition of WTC7 is not available from the existing literature, a gravity area load of 4 kN/m², which represents the mean dead and live loads used in current construction practice, was applied directly to each floor in the model. The boundary condition was pin-supported at the ground column.

5.7.3 Fire Load Simulation

The fire load was applied in Abaqus® as the fire temperature amplitude on the structural members, based on the parametric time–temperature curve (as shown in Figure 5.3) outlined in the Eurocode 1 (BSI, 2002a).

According to NIST NCSTAR (2008), the debris from WTC1 caused structural damage to the southwest exterior, primarily between floors 7 and 17. The fire ignited at least 10 floors; however, fire lasted only on floors 7–9 and 11–13 until the building's collapse. In order to investigate the real behaviour of WTC7 in fire, in the simulation, the fire was set on the southeast part of floor 17 (Figure 5.20) to simulate fire duration in practice. As the purpose of this case study is to demonstrate how to conduct a structural fire analysis, only the fire on floor 17 was simulated.

5.7.4 Calculation of Temperature Increase of Slabs and Structural Steel Members

As mentioned earlier, using the formula from BS EN 1993-1-2 (BSI, 2005a) and BS EN 1994-1-2 (BSI, 2005b) give the formula to work out the increase of temperature for both internal unprotected and

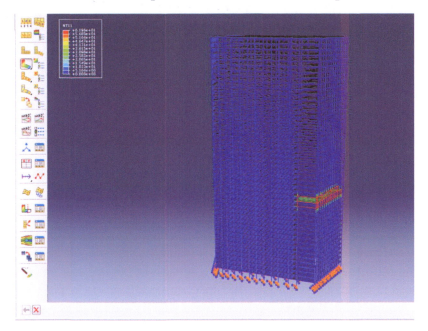

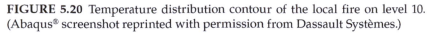

FIGURE 5.20 Temperature distribution contour of the local fire on level 10. (Abaqus® screenshot reprinted with permission from Dassault Systèmes.)

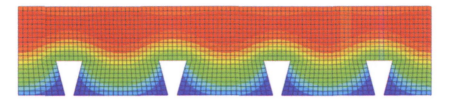

FIGURE 5.21 Distribution of temperature of slab after heat transfer analysis. (Abaqus® screenshot reprinted with permission from Dassault Systèmes.)

protected steelwork were worked out and was applied to the structural members.

To simplify the analysis, the unified temperature was applied to the beams and columns throughout the whole depth by using the formula provided by the codes.

For the slabs, a linear through depth temperature gradient was obtained through the heat transfer analysis in Abaqus® where the temperatures along the thickness of the slab are determined using a parametric fire temperature curve. The result can be seen in Figure 5.21. After the heat transfer analysis, readers can chose five nodes along the depth of the slab and extract the temperature readings and applied to the slabs in the model for the structural fire analysis of the buildings.

5.7.5 Major Abaqus® Command

Rather than using a CAE file, the model here is set up using an INP file. The geometry of the building is first set up using the ETABS program, and then a program developed by Fu (2009) is used to transfer the model from ETABS to INP files in Abaqus®.

The reader can also do a manual setup of the INP file, which is introduced here. The analysis program consists of the below major parts:

1. Define the node coordinates (determine the coordinate of the building using the below command).

```
*node,nset = Nodes
1,88.43674,40.97325,3.6576
2,10.87822,27.55609,3.6576
3,46.51515,27.6018,3.6576
4,70.64726,21.4038,3.6576
5,46.50189,19.77983,3.6576
6,26.80891,19.72235,3.6576
. . . . . . . .
*ncopy,old set = Nodes,new set = DECKlevel, change
  number = 100000,shift
0.,0.,0.25
0
```

2. Define the shell elements.

```
*element,type = s4r,elset = DECK1(The first com-
   mand line defines the shell element type and
   slab name)
100001,111537,111538,111675,111536(The second line
   defines the shell name and four notes for each
   shell.)
. . . . . . . .
```

3. Define the boundary condition (choose all the nodes at the bottom and name them bottomnode).

```
*nset,nset = bottomnode
328
327
326
325
. . . . . .
*boundary
bottomnode,1,6
```

4. Define the frame element (define the beam, column, and bracing).

```
*element,type = b31,elset = W27X1021
1,11444,11445
*beam section, section = I,
elset = W27X1021,material = steel
.28834,.68834,.254,.254,.021082,.021082,.013081
-1.30550171186997E-03,.999999147832277,0
*element,type = b31,elset = W27X1022
2,11446,11447
*beam section,section = I,
elset = W27X1022,material = steel
.28834,.68834,.254,.254,.021082,.021082,.013081
6.83238347633547E-03,.999976658995615,0
. . . . . . . .
```

5. Connect the beam to the slab (use the constraint equation to connect the slab to the beam to make a composite action).

```
*nset,nset = allbeam,elset = allbeam
*ncopy,old set = allbeam,new set = sbeam,change
   number = 100000,shift
0.,0.,0. 25
0.
```

```
*mpc (constraint equation is defined here)
beam,sbeam,allbeam
```

6. Release definition.

```
*elset,elset = momrels1
12061
. . . . . .
*elset,elset = momrels2
12059
. . . . . .
*release
momrels1,s1,M1-M2
momrels2,s2,M1-M2
```

7. Define the fire-protected steel members (make an element set for these steel members with fire protections).

```
*Elset, elset = probeam
6099, 6100, 6101, 6102, 6103, 6104, 6105, 6106,
   6107, 6108, 6109, 6110, 6111, 6112, 6113, 6114
```

8. Define the materials for the steel and concrete.
   a. Define the steel material (this sets up the material property for the steel member under elevated temperatures according to BS EN 1994-1-2: Eurocode 4 [BSI, 2005b]).

```
*MATERIAL,name = Steel
*elastic,type = iso
2.10E+11,0.3,20
2.10E+11,0.3,100
1.89E+11,0.3,200
1.68E+11,0.3,300
1.47E+11,0.3,400
1.26E+11,0.3,500
. . . . . .
*plastic
355e6,0,20
355e6,1.69E-03,20
355e6,0.00E+00,100
355e6,1.69E-03,100
286.49e6,0.00E+00,200
286.49e6,1.52E-03,200
315.59e6,3.36E-03,200
. . . . . .
*EXPANSION
1.23E-5,20.
```

```
1.23E-5,50.
1.25E-5,100.
1.27E-5,150.
1.29E-5,200.
```

   b. Define the concrete material (this sets up the material
      property for the concrete member under elevated tem-
      peratures according to BS EN 1994-1-2: Eurocode 4 [BSI,
      2005b]).

```
*material,name = c60
*elastic
14331210191,0.2,20
11172000000,0.2,100
8621550591,0.2,200
6259090909,0.2,300
4457734077,0.2,400
2828907270,0.2,500
1616689021,0.2,600
962882096,0.2,700
464907252,0.2,800
239715892,0.2,900
119857946,0.2,1000
29964487,0.2,1100
*Concrete Damaged Plasticity
30,,1.16,,0.
*Concrete Compression Hardening
28662420 ,   0        ,, 20
30000000 ,   0.0005 ,, 20
```

9. Define the analysis steps. Two analysis steps are defined
   here. The first is the static step, which is applied to normal
   gravity loads, such as dead and live loads, of the structure.
   The second is the fire analysis step. The static step was
   introduced in Chapter 2; therefore, only the second step is
   demonstrated here.

   In the fire analysis step, the increase of temperature of the
   slab (from the heat transfer analysis introduced in Section
   5.7.4, five node temperatures are chosen) and the protected
   or unprotected beam and columns (calculated using the
   formula from BS EN 1993-1-2: Eurocode 3 [BSI, 2005a]) is
   calculated every 10 min along the fire temperature curve
   (introduced in Section 5.2.1) and lasts until the end of the
   fire temperature curve. Therefore, if you are running a 2 h

fire analysis, there will be 120/10 = 12 steps; only one step is demonstrated here.

```
****Step2**********
*step,unsymm = yes,nlgeom = yes,inc = 5000
*static
60,600,1e-30,60
*temperature
slab-fire,20.1014,163.625,700.431,37.2394,22.0434
  (defining the temperature of the slabs,
  temperature of 5 nodes are selected from the
  heat transferring analysis)
**protected beam (defining the temperature of the
  protected beam and column)
probeam,48.3490677963654
procolumn,49.2077087054519
**unprotected beam (defining the temperature of
  the unprotected beam and column)
beam-fire,60.5778721012576
column-fire,46.0727787944676
*end step
```

5.7.6 Modelling Results Interpretation

The results of the finite-element analyses undertaken using the above inputs are outlined here.

The results presented consist of contour plots (Figure 5.20) across the compartment at the peak response of the structure. Figure 5.20 shows the temperature distribution at the floor where the fire is set. Readers can also plot the contour for any particular time they want to investigate, such as at the end of the analysis, and also the time history of certain parameters.

To plot the time history of a parameter you want to investigate, use the procedure shown in Chapter 2:

- Go to *Result* and click on *XY Date*.
- A new window will pop up.
- Click on *ODB Field Output*.
- A new window will pop up (Figure 5.22).

Select the result parameters you want to investigate, such as displacement.

- Choose the node you want to investigate (Figure 5.23) from the model.

Then the vertical displacement of that node can be plotted.

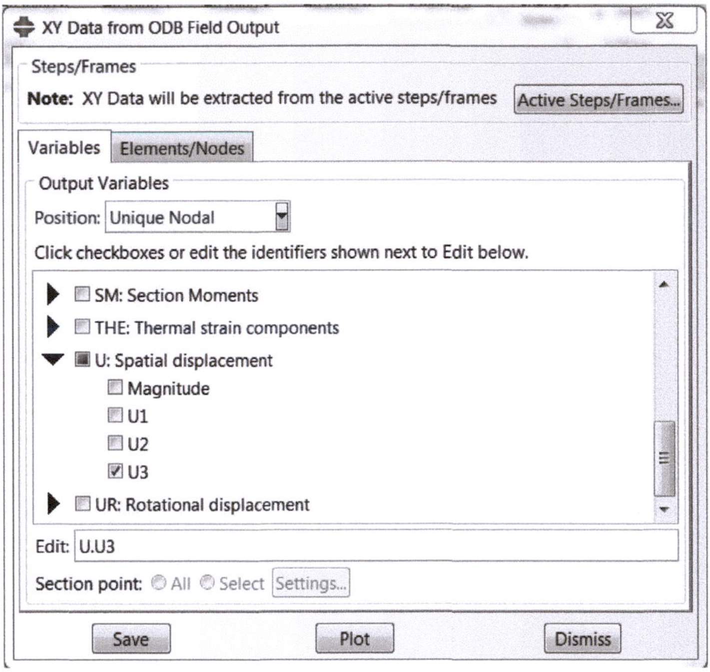

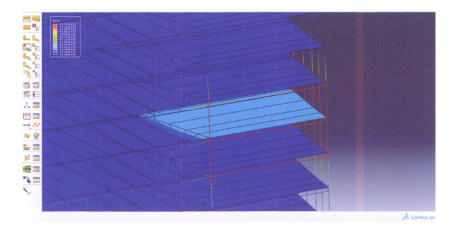

FIGURE 5.22 Vertical displacement (U3) is selected. (Abaqus® screenshot reprinted with permission from Dassault Systèmes.)

FIGURE 5.23 A node (highlighted in red) has been selected. (Abaqus® screenshot reprinted with permission from Dassault Systèmes.)

If you want to investigate the internal force of a certain structural member, then

- Click on *ODB Field Output* and select the result section force (Figure 5.24).
- Choose an element (Figure 5.25) from the model.

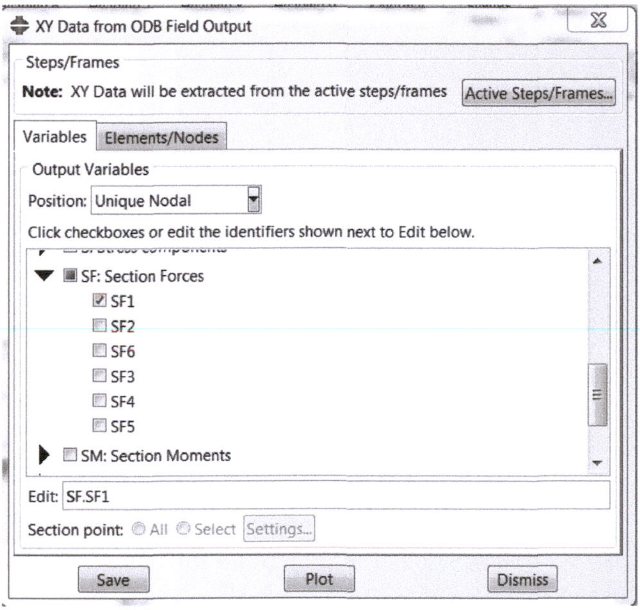

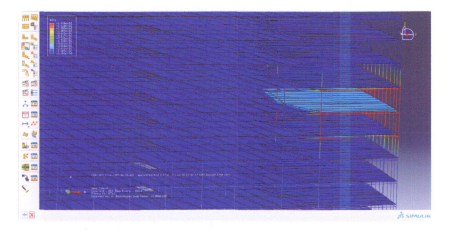

FIGURE 5.24 Axial force (SF1) is selected.

FIGURE 5.25 An element (highlighted) has been selected. (Abaqus® screenshot reprinted with permission from Dassault Systèmes.)

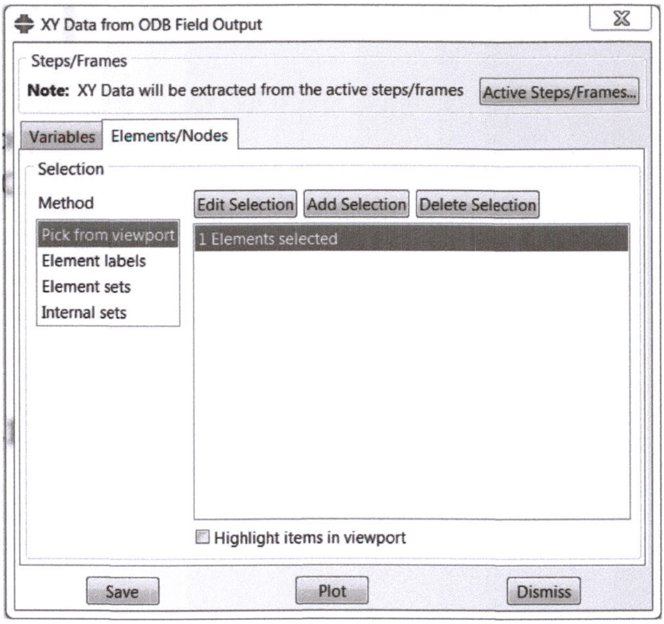

FIGURE 5.26 Selecting element. (Abaqus® screenshot reprinted with permission from Dassault Systèmes.)

- Click on *Done*; select elements will be shown as in Figure 5.26.
- Click on *Plot*; you can get the result of the axial force (since we chose section force SF1, an axial force), which can also be exported to Excel files.
- Go to *Report on the Ribbon* and click on *XY*; a new window will pop up, as shown in Figure 5.27.
- Click on *Setup*; you can define the parameters for the output file.
- Select the XY plot you want to output and click *OK*. A text file will be generated that you can copy and paste into an Excel file for further data plotting, as shown in Figure 5.28.

5.7.7 *Progressive Collapse Potential Check*

Section 5.7.6 introduces how to extract the analysis result. If we want to check the potential of collapse or partial collapse of a building, we first need to check the level of damage of the major structural elements, such as the columns and braces, through the plotting of the axial force or plastic strain developed during the analysis. If any failure of the structural members is noticed, we should remove

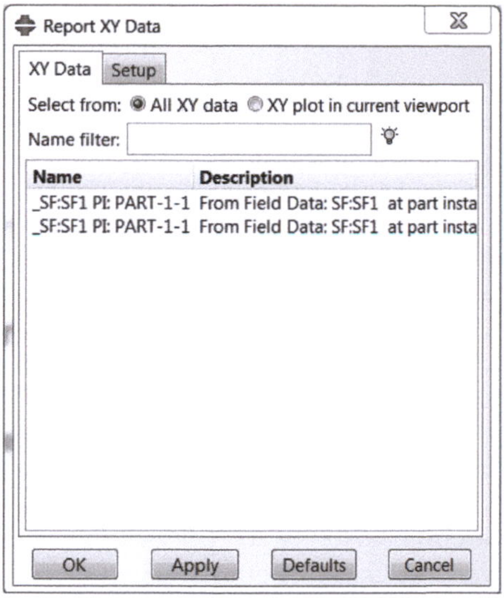

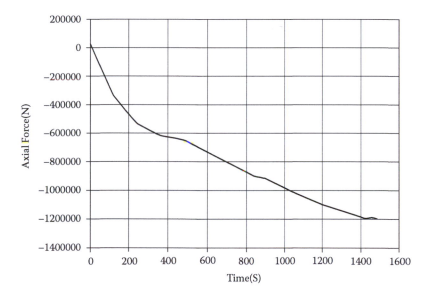

FIGURE 5.27 Generating text file. (Abaqus® screenshot reprinted with permission from Dassault Systèmes.)

FIGURE 5.28 Axial force of selected members in Excel. (Abaqus® screenshot reprinted with permission from Dassault Systèmes.)

those members from the model and follow the modelling procedure introduced in Chapter 2; a progressive collapse analysis can be performed. Readers can refer to Chapter 2 for detailed progressive collapse analysis procedures.

References

Bailey, C.G., and Moore, D.B. 2000. The structural behaviour of steel frames with composite floor slabs subject to fire. Part 2: Design. *Structural Engineer*, 78(11), 29–33.

British Steel, Swinden Technology Centre. 1999. The behaviour of multistorey steel framed buildings in fire. Rotherham, UK: British Steel, Swinden Technology Centre.

BSI (British Standards Institution). 1987. Incorporating amendment no. 1. Fire tests on building materials and structures. Part 20: Method for determination of the fire resistance of elements of construction (general principles). BS 476-20. London: BSI.

BSI (British Standards Institution). 1990. Structural use of steelwork in buildings. Part 8: Code of practice for fire resistant design. BS 5950. London: BSI.

BSI (British Standards Institution). 2002a. Eurocode 1: Actions on Structures. Part 1-2: General actions—Actions on structures exposed to fire. BS EN 1991-1-2. London: BSI.

BSI (British Standards Institution). 2004. Eurocode 2: Design of concrete structures. Part 1-2: General rules—Structural fire design. BS EN 1992-1-2. London: BSI.

BSI (British Standards Institution). 2005a. Eurocode 3: Design of steel structures. Part 1-2: General rules—Structural fire design. BS EN 1993-1-2. London: BSI.

BSI (British Standards Institution). 2005b. Eurocode 4: Design of composite steel and concrete structures. Part 1-2: General rules. Structural fire design. BS EN 1994-1-2. London: BSI.

CSI (Computer and Structures). 2008. *ETABS theory manual*. Version 9.2. New York: Computer and Structures.

Dave, P. 2005. Madrid tower designer blames missing fire protection for collapse. *New Civil Engineer*, June 2.

Fu, F. 2009. Progressive collapse analysis of high-rise building with 3-D finite element modeling method. *Journal of Constructional Steel Research*, 65(6), 1269–1278.

Fu, F. 2015. Three-dimensional finite element analysis of the whole-building response of a steel-composite tall building in fire. Under review.

Institution of Structural Engineers. 2007. *Guide to the Advanced Fire Safety Engineering of Structures*. London: Institution of Structural Engineers, August.

ISO (International Organization for Standardization). 1999. Fire-resistance tests—Elements of building construction. Part 1: General requirements. ISO 834-1. Geneva: ISO.

Jacobs, A. 2009. China TV network apologizes for fire. *New York Times*, February 10.

NILIM (National Institute for Land and Infrastructure Management). 2005. Report on the Windsor Building fire in Madrid, Spain [in Japanese]. Tsukuba, Japan: NILIM, July 1.

NIST (National Institute of Standards and Technology) NCSTAR (National Construction Safety Team). 2008. Federal building and fire safety investigation of the World Trade Center disaster. Final report on the collapse of World Trade Center Building 7. Gaithersburg, MD: NIST, U.S. Department of Commerce.

Talamona, D., Franssen, J.M., Schleich, J.B., and Kruppa, J. 1997. Stability of steel columns in case of fire: Numerical modeling. *Journal of Structural Engineering*, 123(6), 713–720.

University of Manchester. 2005. The Windsor Tower fire, Madrid. http://www.mace.manchester.ac.uk/project/research/structures/strucfire/CaseStudy/HistoricFires/BuildingFires/default.htm.

Wang, Y. 2004. Postbuckling behavior of axially restrained and axially loaded steel columns under fire conditions. *Journal of Structural Engineering*, 130(3), 371–380.

Wang, Y.C. 2000. An analysis of the global structural behaviour of the Cardington steel-framed building during the two BRE fire tests. *Engineering Structures*, 22(5), 401–412.

Yang, D.B., Zhang, Y.G., and Wu, J.Z. 2010. Elasto-plastic buckling analysis of space truss structures with member equivalent imperfections considered using Abaqus®. Presented at Proceedings of the Third International Conference on Modelling and Simulation, Wuxi, People's Republic of China, June 4–6.

CHAPTER **6**

Design and Analysis of Buildings under Blast Loading

6.1 Introduction

Among the different incidents that result in progressive collapse, an explosion is one of the major reasons. An explosion can cause damage to the building's structural frames, which may cause partial or full collapse of the structure. The partial collapse of the Alfred P. Murrah Federal Building in Oklahoma City due to the car bomb attack in 1995 (ODCEM, 1995) is a famous example of a blast-triggered building collapse.

In order to prevent partial or full collapse of structures, the design of buildings against blast loading and relevant special design requirements of buildings against blast load become increasingly important in design consideration. Depending on the category of the buildings, a design check for buildings under blast load is also required by statute in the UK.

Therefore, in this chapter, collapse incidents around the world triggered by a blast load are introduced, the fundamentals of blast loading are presented, and the design methods of blast-resistant structural elements and building for progressive collapse resistance are introduced as well. At the end of the chapter, a modelling example for blast analysis of the Murrah Federal Building is demonstrated using the finite-element package Abaqus®.

6.2 Blast-Induced Progressive Collapse Incidents around the World

The collapse of a building due to an explosion or impact loading is not rare. High explosives can partially or totally damage a building. In this section, several collapse incidents caused by blast loading are discussed, and the collapse mechanisms of the structures are introduced.

6.2.1 Alfred P. Murrah Federal Building Collapse

The partial collapse of the Alfred P. Murrah Federal Building in Oklahoma City, Oklahoma, is a famous example of a disproportionate collapse of a building caused by blast loading. In 1995, the collapse was caused by a bomb equivalent to 1800 kg of TNT. It detonated in a rental truck parked in a loading lane within a distance of about 4.3 m on the north side of the Murrah Building. The larger explosion pressure destroyed approximately one-third of the Murrah Building (Figure 6.1). The entire north face of the structure collapsed, and the rest of structure received extensive damage (ODCEM, 1995).

FIGURE 6.1 Murrah Federal Building bombing. (From http://www.defense imagery.mil/imagery.html. This image is the work of a U.S. military or Department of Defense employee, taken or made as part of that person's official duties. As a work of the U.S. federal government, the image is in the public domain.)

The Alfred P. Murrah Federal Building was a reinforced concrete structure with a nine-storey office building, two one-storey wings, and a multilevel car park. The major lateral stability of the building was provided by a moment frame working together with a cast-in-place concrete core on the south side (Figure 6.2). The east and west elevated structures also contained prefabricated spandrels. To provide street access to the first floor, the second floor was held back; therefore, on the third floor, a large transfer beam, supported by two-storey-high columns spaced 12.3 m over centre (Figure 6.3), carried all of the load from level 3 up. This arrangement made the building vulnerable to the effects of an explosion, as this transfer beam was the key element supporting the gravity load of the upper levels.

Figure 6.3 also shows that one column was destroyed directly by the explosion, which was the column nearest to the detonation position. In addition, the large blast wave caused the shear failure to its adjacent two columns. Therefore, the transfer beam lost support due to the failure of these three columns, and the excessive loading from level 3 and above caused the failure of the transfer beam. This transfer beam was the key element for supporting the gravity load from level 3 up; therefore, its failure caused the progressive failure of the structures in the north face.

6.2.2 Argentine Israeli Mutual Association Bombing (Whole Building Collapse), Buenos Aires, Argentina

In 1994, a suicide bomber drove a van bomb loaded with about 275 kg of explosive mixture into the Jewish Community Centre Building. The exterior walls of this five-storey building were constructed of brick masonry, which supported the floor slabs. The air blast from the bomb virtually destroyed the whole building. The exterior walls were demolished completely, which led to progressive failure of the floor slabs, and therefore the total collapse of the building. This is an example of how localised damage leads to whole building collapse.

6.2.3 Brighton Hotel Bombing, UK (Partial Collapse)

A bomb made of 9 kg of Frangex was placed in one of the rooms in the hotel. The device was detonated by a timer on October 12, 1984. The middle section of the building collapsed into the basement, leaving a gaping hole in the hotel's façade.

Global collapse was not triggered because of the strong robustness of this type of Victorian hotel. It has closely spaced columns, smaller beam span, and strong beam-to-column connections, which make it less vulnerable when a local collapse occurs. The rest of the

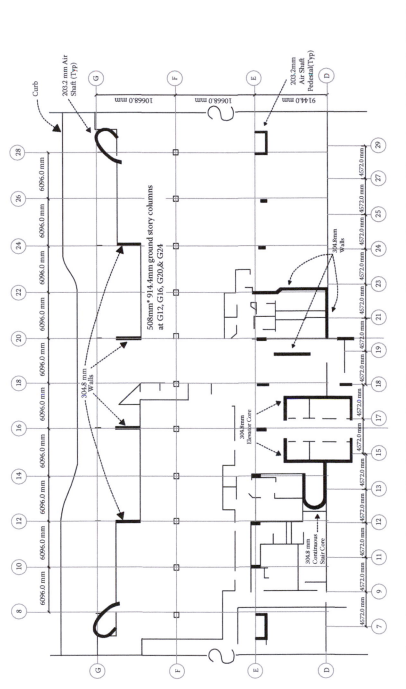

FIGURE 6.2 General Agreement of ground level. (Reproduced based on an architectural drawing of Shaw Associates/Locke Wright Foster for the General Service Administration.)

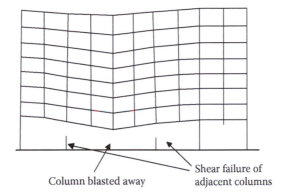

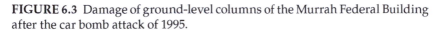

Column blasted away   Shear failure of
                        adjacent columns

FIGURE 6.3 Damage of ground-level columns of the Murrah Federal Building after the car bomb attack of 1995.

structure remained standing, as the beams are tied strongly to the columns, and therefore disproportionate collapse was not triggered.

6.3 Basics of Blast Loading

To understand the collapse mechanisms of buildings, it is imperative to understand the fundamentals of blast loading, such as blast load characteristics, the scaling law principle, blast profiles, and dynamic increase factors. In this section, this basic knowledge is discussed in detail.

6.3.1 Explosion and Blast Load

An explosion is a large-scale, rapid, and sudden release of energy. Explosive materials can be classified as solids, liquids, or gases. It can also be identified as a physical, nuclear, or chemical explosion. The detonation of a condensed high explosive generates hot gases under pressures of up to 10–30 GPa and temperatures of about 3000°C–4000°C. The blast effects of an explosion are in the form of a shock wave composed of a high-pressure shock front. The blast wave instantaneously increases to a value of pressure above the ambient atmospheric pressure; after that, it starts to drop quickly. The blast wave expands outward from the centre of the detonation, with maximum overpressures decaying with distance. The time history of blast pressure development in free air is shown in Figure 6.4.

In real situations, a blast wave may impinge on a solid surface of a building (or on a dense medium) and be reflected. These reflections, particularly in built-up areas, can create complex loading conditions.

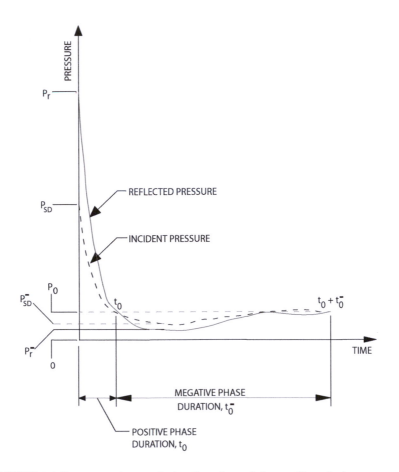

FIGURE 6.4 Pressure–time variation for a free-air burst. (Permission to reproduce and derive from Figure 2-5 of UFC 3-340-02, Unified Facilities Criteria (UFC), Structure to resist the effects of accidental explosions, 2008, Change 2, WBDG ®Whole Building Design Guide, a program of the National Institute of Building Sciences)

The actual blast loads can be either reduced or enhanced due to the presence of other buildings. Therefore, the above factors should also be included in the blast overpressure profile. A reflected blast wave can be used as shown in Figure 6.4. However, this makes predicting blast loading much more complicated, so reasonable simplifications need to be made.

The types of explosion can be classified as unconfined explosions, confined explosions, and explosives attached to a structure.

Depending on the height of the location of detonation, for unconfined explosions, it can be further subdivided into free-air burst, air burst, and surface burst.

1. A free-air burst occurs in free air high above ground level. Therefore, there is no amplification of the blast waves prior to contact with the structure.
2. An air-burst explosion also occurs above ground level; however, intermediate amplification of the wave caused by ground reflections occurs prior to the arrival of the initial blast wave at a building.
3. A surface-burst explosion occurs when the detonation is close to or on the ground surface. The initial shock wave is reflected and amplified by the ground surface to produce a reflected wave.

6.3.2 Blast Wave Scaling Laws and Simplified Blast Load Profile

In design practice, a convenient way to represent blast wave parameters is to plot them against scaled distance Z. The scaled distance is used widely to determine blast-wave characteristics as shown in Figure 6.4. Both incident and reflected blast wave parameters may be represented in this manner. Therefore, the blast loadings are evaluated using empirical relationships based on the scaling law principle by design guidances such as SCI (Yandzio and Gough, 1999) and UFC 3-340-02 (UFC, 2008).

The scaling law principle, formulated independently by Hopkinson (1915) and Cranz (1926), is used extensively to determine blast wave characteristics. It is based on the conservation of momentum and geometric similarity. The empirical relationship is described as a cube-root scaling law and is defined as

$$Z = \frac{R}{W^{1/3}} \qquad (6.2)$$

where Z is the scaled distance ($m/kg^{-1/3}$), R is the range from the centre of the charge, and W is the mass of the spherical TNT charge (kg). When Z is determined, the characteristics of the blast wave such as the peak overpressure can be therefore worked out by checking Figure 6.4B. The reader can refer to UFC 3-340-02 (UFC, 2008) for further information. The progression of a free-air burst is best represented by a pressure–time history curve. The magnitude of the pressure caused by a blast wave is usually quoted as an overpressure (the pressure increase relative to ambient pressure). The characteristics of the blast wave used for calculation purposes are shown as a blast overpressure–time curve. It is usually adequate to assume that the growth and decay of blast overpressure is linear. For the positive

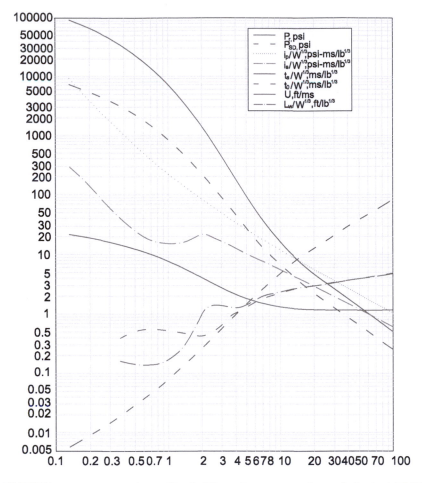

FIGURE 6.5 Positive Phase Shock Wave Parameters for a Spherical TNT Explosion in Free Air at Sea Level. (Permission to reproduce and derive from Figure 2-7 of UFC 3-340-02, Unified Facilities Criteria (UFC), Structures to resist the effects of accidental explosions, 2008, Change 2, WBDG ®Whole Building Design Guide, a program of the National Institute of Building Sciences)

overpressure phase, a simplification is made where the impulse of the positive phase of the blast is preserved and the decay of overpressure is assumed to be linear. This simplification is shown in Figure 6.6. As introduced in Section 6.4, when designing a structural element under blast loading, a designer can directly use the idealization of blast loading.

6.3.3 Material Behaviours at High Strain Rates
Under the blast load, the structural materials exhibit different behaviours due to the high rates of strain (in the range of 10^2–10^4/s).

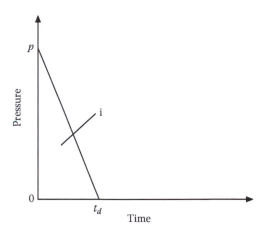

FIGURE 6.6 Idealization of blast loading in the design.

The dynamic ultimate strength increases, and it can be much greater than the static ultimate strength. The mechanical properties of the structural steels under blast loading are noticeably affected by the rate at which straining takes place.

The dynamic design yield stress of the steel $F_{y,des}$ for bending is given by SCI (Yandzio and Gough, 1999):

$$F_{y,des} = a(DIF)\, F_y \qquad (6.1)$$

where a is a factor that takes into account the fact that the yield stress of a structural component is generally higher than the minimum specified value; for S275 and S355 steels, $a = 1.10$. *DIF* is the dynamic increase factor for structural steels; it can be checked in Table 9.1 of SCI (Yandzio and Gough, 1999).

Similarly, the dynamic strength of reinforced concrete can be worked out following a similar equation, and the dynamic increase factors for concrete can be checked in Table 9.3 of SCI (Yandzio and Gough, 1999).

6.3.4 Response Regimes of Structural Elements

Under blast loading, structures behave dynamically. The response of a structure (or a structural element) is determined greatly by the ratio between its natural period and the duration of the blast. According to SCI Publication 244 (Yandzio and Gough, 1999), three response regimes are defined, which are based on the natural period of the structure:

Impulsive $\qquad t_d/T < 0.4$

Dynamic $\qquad 0.4 < t_d/T < 2$

Quasi-static $\qquad t_d/T > 2$

where t_d is the duration of the blast load and T is the natural period of vibration of the structural element.

It is important to note that for a particular blast wave, the response could be impulsive for one structure but quasi-static for another because of the different natural periods of vibration of each structure.

6.4 Design of Buildings under Blast Loading

When designing a building to resist blast loading, several design guidances are available internationally. UK SCI Publication 244 (Yandzio and Gough, 1999) provides guidance on the design of commercial and public buildings where there is a requirement to provide protection against the effects of explosions caused by the detonation of explosives. A philosophy for the design of buildings to reduce the effects of attack is introduced, and a design procedure is proposed. The robustness of buildings and the prevention of disproportionate collapse are also discussed.

In the United States, the Army technical manual, UFC 3-340-02 (UFC, 2008), is one of the most detailed U.S. design guidances for introducing the design of buildings against blast loading. FEMA 427 (FEMA, 2003) provides design measurements to reduce physical damage to the structural and nonstructural components of buildings and related infrastructures during conventional bomb attacks, as well as attacks using chemical, biological, and radiological (CBR) agents.

6.4.1 Explosion Scenarios

When designing a building under blast loading, it is imperative to understand the major explosion scenarios. There are many ways in which an explosive device may deliver an attack, such as vehicle bombs, package bombs, mortar bombs, culvert bombs, and incendiary devices. For detailed information, refer to Table 6.1 in SCI (Yandzio and Gough, 1999).

6.4.2 Iso-Damage Diagrams (Pressure–Impulse Diagrams)

From the previous sections, it can be seen that the design of a structure against explosion is a complicated procedure. One needs to assess the structural response mathematically from the first principles. In real design practice, engineers can design buildings under blast loading using analytically derived pressure–impulse diagrams (or iso-damage curves), which are derived from experimental evidence or real-life events. These diagrams can readily predict levels of damage for various load–impulse combinations. They cover a whole range of possible response regimes, from quasi-static to impulsive regimes. They provide an effective method to relate a specific damage level to a combination of blast pressures and impulses imposed on a particular structural element.

The first pressure–impulse curves (or isodamage curves) were derived from a study of houses damaged by bombs dropped on the UK in the Second World War (Jarrett, 1968). The results of such investigations are used in the evaluation of safe standoff distances for explosive storage in the UK. When blast loading is applied, different levels of damage are inflicted on the buildings. Jarrett (1968) gives detailed classifications of different levels of damage to structures, which are the basis of the iso-damage curve. They are defined as Categories A, B, C_b, C, and D, where the most severe damage is in Category A, with the buildings being completely demolished.

Figure 6.7 is an isodamage curve of the above damage levels, where the axes of the curve simply represent peak overpressure versus specific impulse.

In design practice, pressure–impulse diagrams are used together with blast parameter and scaled distance graphs by simply overlaying them into one diagram (Figure 6.7). This allows the development of equations to describe specific damage levels. By superimposing the blast parameters, such as charge mass and scaled distance, the damage to buildings caused by specific explosive devices can be assessed in Figure 6.7. For example, it shows that for 10 tonnes of TNT at the 100 m range, the damage level falls into Category A, which indicates almost complete demolition of a building.

6.4.3 Human Response to Blast Loading and Survival Curves

In addition to predicting the damage to buildings, in the design process, it is imperative to understand the level of human injury to specific blast attacks. The human response to blast loading can be checked using pressure impulse diagrams from UFC 3-340-02 (UFC, 2008). The correspondent survival curves for humans can also be checked in UFC 3-340-02 (UFC, 2008).

It was found that the orientation of a person (standing, sitting, prone, face on or side on to the pressure front) relative to the blast

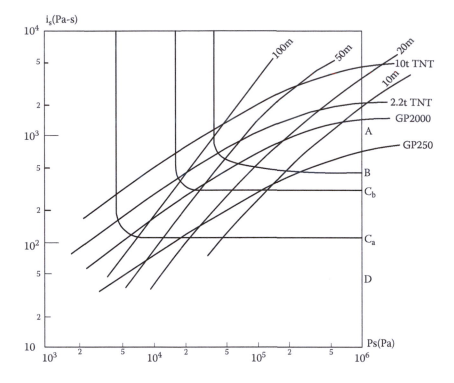

FIGURE 6.7 Pressure–impulse diagram for damage to houses with a range–charge weight overlay. (From Hetherington and Smith, *Blast and Ballistic Loading of Structures*, Boca Raton, FL: CRC Press, 1994. Copyright © CRC Press.)

front and the shape of the pressure front (fast or slow rise, stepped loading) are significant factors in determining the amount of injury sustained.

There are three levels of injury (Baker et al., 1983):

- Primary injury
- Secondary injury
- Tertiary injury

Primary injury is due directly to blast wave overpressure and duration, which can be combined to form a specific impulse. The result also depends on a person's size, gender, and age. The most likely organs to be damaged include the lungs, which are prone to haemorrhage and oedema; the ears (particularly the middle ear), which can rupture; the larynx; the trachea; and the abdominal cavity. Tests have indicated that the air-containing tissues of the lungs are the critical target organs in blast pressure injuries. The release of air bubbles from disrupted alveoli of the lungs into the vascular system probably accounts for most deaths. An estimation of man's lung and ear response to blast pressure is presented in Figures 1-2 and 1-3 in UFC 3-340-02 (UFC, 2008), and they are reproduced here in Figures 6.8 and 6.9.

Secondary injury is due to impact by missiles (e.g., fragments from weapon casing) created by explosive devices. Such missiles

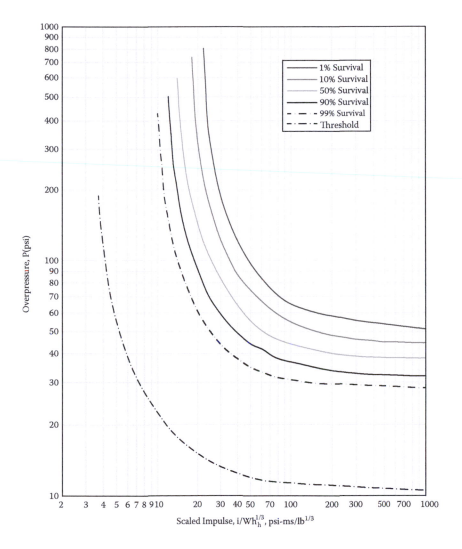

FIGURE 6.8 Survival curves for lung damage. Wh, weight of human being (lb). (From UFC, Structures to resist the effects of accidental explosions, 2008 change 2, UFC 3-340-02, Washington, DC: National Technical Information Service, 2008, Figure 1-2. With permission from Whole Building Design Guide® (WBDG), a program of the National Institute of Building Sciences.)

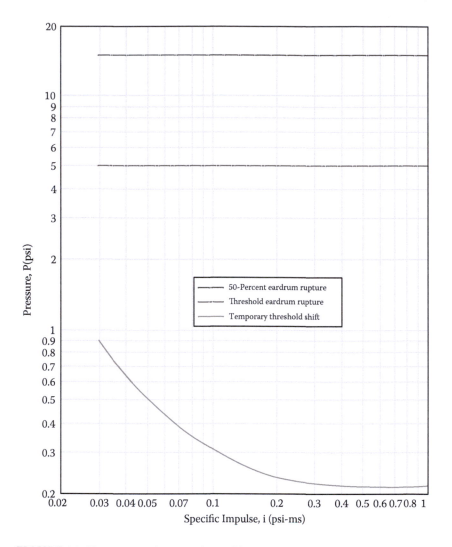

FIGURE 6.9 Human ear damage due to blast pressure. (From UFC, Structures to resist the effects of accidental explosions, 2008 change 2, UFC 3-340-02, Washington, DC: National Technical Information Service, 2008, Figure 1-3. With permission from Whole Building Design Guide® (WBDG), a program of the National Institute of Building Sciences.)

produce lacerations, penetration, and blunt trauma (a severe form of bruising).

Tertiary injury is due to displacement of the entire body, which is inevitably followed by high declarative impact loading, where, of course, most of the damage occurs.

6.5 Structural Design to Prevent Collapse of Buildings under Blast Loading

In this section, the methods of structural analysis and design guidance to resist blast loads are introduced. It consists of designing blast-resistant structural elements and designing a building to prevent disproportionate collapse under blast loading.

6.5.1 Acceptance Criteria

The criteria used to assess the performance of a structure subjected to blast loading are strength limit and deformation limit. Where strength governs design, failure is defined as occurring when the design load or load effects exceed the design strength. In addition, deformation limits are frequently used in the design. They are introduced in Section 6.5.2.

6.5.2 Design of Blast-Resistant Steel or Concrete Elements

Blast-resistant structural elements are essential in the design of buildings to prevent disproportionate collapse. Therefore, in this section, detailed guidance on how to design blast-resistant structural elements is introduced.

The controlling criterion in the design of blast-resistant structural elements is normally a limit on the deformation or deflection of the element. In this way, the degree of damage sustained by the element may be controlled. The damage level that may be tolerated in any particular situation will depend on what is to be protected, for example, the structure itself, the occupants of a building, or the equipment within the building.

Protection is divided into two major categories (as shown in Table 6.1):

Category 1: Protection of personnel and equipment through the attenuation of blast pressures, shielding them from the effects of primary and secondary fragments and falling portions of the structure

Category 2: Protection of the structural members themselves from collapse under the action of blast loading

It should be noted that these limits imply extensive deformation of the elements and the need for subsequent repair or replacement before being reused.

Table 6.1 Summary of Design Requirements for Steel and Concrete Structures

	Protection Category			
	1		2	
	θ	μ	θ	μ
Reinforced concrete beams and slabs	$2°$[a]	N/A	$4°$[b]	N/A
Structural steel beams and plates[c]	$2°$	10	$12°$	20
Steel–concrete–steel composite	$2°$	N/A	$5°$	N/A

Source: Modified based on Tables 6.3, 7.3, and 8.2 of Cormie et al., *Blast Effects on Buildings*, 2nd ed., London: Thomas Telford, 2009.

[a] Shear reinforcement in the form of open or closed "blast links" must be provided in slabs for $\theta > 1°$. Close links (shape code 63 in BS 8666 [BSI, 2005]) must be provided in all beams.

[b] Support rotations of up to $8°$ may be permitted when the element has sufficient lateral restraint to develop tensile membrane action. Further guidance regarding the tensile membrane capacity of reinforced concrete slabs may be found in UFC 3-340-02 (UFC, 2008).

[c] Adequate bracing must be provided to ensure the corresponding level of ductile behaviour.

There are two methods to limit the element deformations:

1. Support rotation, θ. In the design, engineers can check the rotation of the structural element against required deformation limit tables, as shown in Table 6.1.
2. Ductility ratio, μ. In the design, engineers can check the ductility ratio of the structural element against required deformation limit tables (e.g., Table 6.1). The method to calculate the ductility ratio is as follows:

$$\mu = \frac{total \text{ deflection}}{\text{deflection at elastic limit}} \tag{6.3}$$

6.5.3 Summary of Procedures for Designing Blast-Resistant Steel or Concrete Members

A simple design procedure is presented here showing how to determine the response and adequacy of an individual structural member subjected to a blast load. The response of the member is based on the single-degree-of-freedom analysis method.

1. Determine the blast load characteristics assuming that the blast load is triangular in profile and that the rise time is zero (Figure 6.5).

2. Assume a loading or response regime, that is, impulsive or dynamic/quasi-static.
3. Determine the dynamic material properties based on DIFs.
4. Assume an acceptable response criterion based on a maximum allowable ductility ratio, μ, or support rotation, θ.
5. Estimate the maximum member resistance value, R_m. Choose a preliminary element size for steel or concrete members. Select a steel section size that is not prone to lateral buckling.
6. Determine the required value of the plastic moment of resistance, M_p.
7. Calculate the natural period of the structural member using transformation factors for mass, stiffness, and load from the relationship.
8. Calculate the ductile ratio, μ, which is the required ductility ratio for the member selected under the applied blast load. Check that the ductility ratio is acceptable according to the criterion in Table 6.1. If not, a different section must be selected and the process repeated. In some cases, calculate the support rotation, θ, and check that it is satisfactory.
9. Check that the shear stress is satisfactory.
10. For steel members, check that lateral torsional buckling does not occur. Note that the plastic hinge compressive zone can be quite long. Check that the connections at the ends of the beam are adequate.
11. Check that the correct loading or response regime has been chosen. If the quasi-static case is not appropriate, the procedure has to be repeated using impulsive conditions.

6.5.4 Beam–Column Connections

In the design of buildings under blast loading, beam-to-column connections are particularly important, because when the structure is subjected to blast loading, the connection forces are frequently very large. As mentioned in Chapter 2, membrane tension will bring a large axial force to the connections. Therefore, in the design, the connection should be guaranteed to be able to accommodate the extra force caused by the blast loading.

In addition, under the blast loading, high strain rates will also result in increased risk of brittle fracture. Therefore, good welding procedures are required for steel connections if fractures are to be avoided during overload.

6.5.5 Design Principle for Blast Loading and Measures to Prevent Disproportionate Collapse

According to SCI Publication 244 (Yandzio and Gough, 1999) and UFC 3-340-02 (UFC, 2008), the basic design principle for building against a blast load is to minimize the likelihood and magnitude of threat, which can be achieved by adopting preventive measures that discourage or impede an attack. One example is that the standoff distance of a blast can be increased through using more land and securing more perimeters with barriers; however, this increases the cost.

The building should also be designed to protect the people and assets in it from the effect of blast waves and projectiles by providing the occupants with either a safe area or an effective escape route and assembly area.

The design procedure for the protection of buildings is summarized in the flowchart shown in Figure 6.10.

In the event of fire, design measurements should also prevent it from burning out of control. Detailed guidance was introduced in Chapter 5.

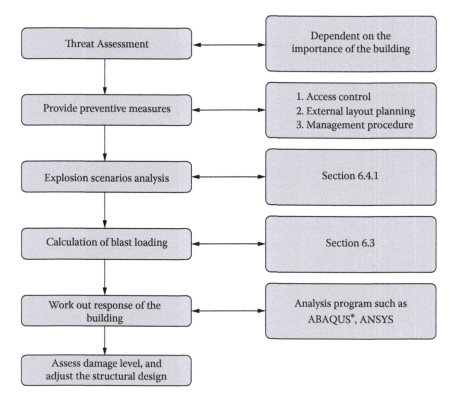

FIGURE 6.10 Flowchart of design procedure for building under blast loading.

In the terms of structural design, the building should also be designed to prevent progressive collapse. This is achieved by providing sufficient ductility and redundancy, such as an alternative load path in the structural design, as mentioned in Chapter 2.

6.6 Modelling Examples of Two-Storey Building under Blast Load Using Abaqus®

In the earlier part of this chapter, the design method for building under blast loading was introduced; however, due to its complexity, it required several manual calculations. Therefore, a computer program is an effective tool to perform blast analysis. In the following two sections, we introduce how to model the blast effect on a building in Abaqus®.

In this section, the modelling techniques of a two-storey building under blast loading are introduced. The model is set up using the three-dimensional (3D) solid element available in Abaqus®. All material properties are used with the consideration of DIFs by simply working out the DIF and multiplying it to the normal material properties.

6.6.1 *Prototype Building*
A two-storey steel and composite building was modelled with Abaqus® (Figure 6.11). The building used a steel beam and column system with composite slabs connected to the beam using shear studs. This is one of the conventional construction projects in current design practice.

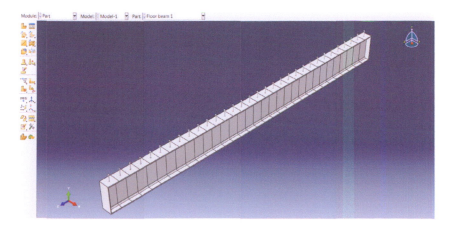

FIGURE 6.11 Part representing steel beam with stud simulated. (Abaqus® screenshot reprinted with permission from Dassault Systèmes.)

The 3D model was set up using Abaqus®, which replicates the true sizes of all the structural elements and the real dimension of the two-storey steel composite building. The major modelling steps are discussed below.

6.6.2 3D Model Setup

- Different parts, such as beam, slab, and columns, were first defined in the *Part* module (Figure 6.11), which demonstrated a beam, with the shear studs simulated as well.
- After all the parts, such as the beams and slabs, are defined, they are assembled in the *Assembly* module (Figure 6.12).
- In the latest version of Abaqus®, you can merge all the parts into a two-storey building, to make sure all the structural members are connected to each other.
- Another option is to define the contact element between different parts representing structural members; however, due to the complexity of the model, this becomes quite difficult.
- The material properties for both steel and concrete members can be defined in the *Property* module (Figure 6.13). However, make sure to increase the material strength by multiplying the dynamic increase factors, as shown in Equation 6.2.

6.6.3 Defining Explosion Step and Blast Loading

In the latest version (6.13) of Abaqus®, the *CONWEP* module for blast loading application has been developed. In this section, how to apply the blast loading is demonstrated.

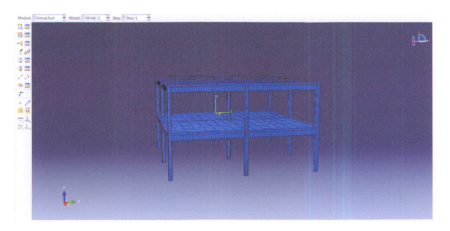

FIGURE 6.12 Assembly into a two-storey building. (Abaqus® screenshot reprinted with permission from Dassault Systèmes.)

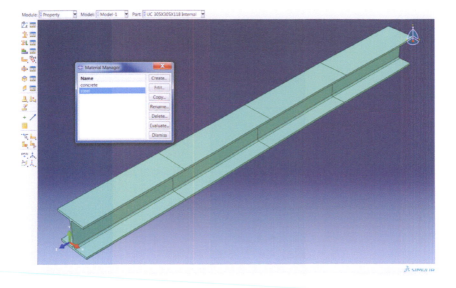

FIGURE 6.13 Defining material properties. (Abaqus® screenshot reprinted with permission from Dassault Systèmes.)

- In the *Step* module, click on *Step Manager*. A new window will pop up (Figure 6.14).
- Click on *Create* and choose *Dynamic, Explicit* procedure. Name it *Explosion* (Figure 6.15).
- In the *Interaction* module, choose *Create Interaction* property. A New window will pop up. Choose *Incident Wave* and give it the name *Blast* (Figure 6.16).
- Click on *Continue*. A new window will pop up (Figure 6.17). Choose *Air Blast*, define the *CONWEP Charge*, and click *OK*.
- Define the reference points. Click on *Create Reference Point*; choose a location or enter the coordinates of the reference points. These reference points can be used to define the location of the explosive detonation.

Step Manager

Name	Procedure	NIgeom	Time
✔ Initial	(Initial)	N/A	N/A

Create... Edit... Replace... Rename... Delete... NIgeom... Dismiss

FIGURE 6.14 Defining explosion step. (Abaqus® screenshot reprinted with permission from Dassault Systèmes.)

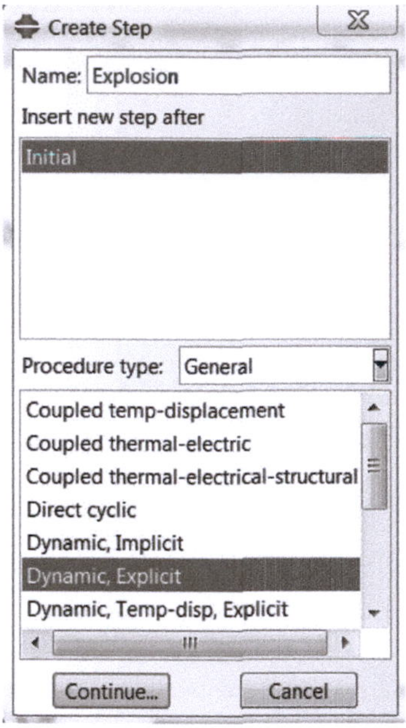

FIGURE 6.15 Defining explosion step. (Abaqus® screenshot reprinted with permission from Dassault Systèmes.)

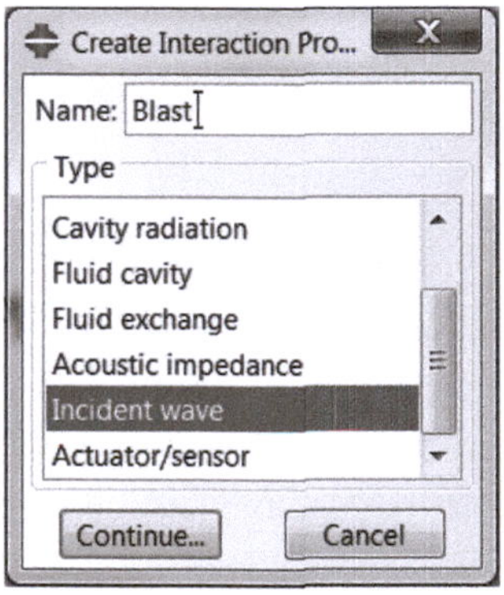

FIGURE 6.16 Defining blast loading. (Abaqus® screenshot reprinted with permission from Dassault Systèmes.)

- In the *Interaction* module, choose *Create Interaction*. A new window will pop up. Choose *Incident Wave* (Figure 6.18).
- It is now required to have a reference point as the resource point. Choose the reference point we defined in the previous step, RP3; this means the blast will be detonated at the base of one of the corner columns.

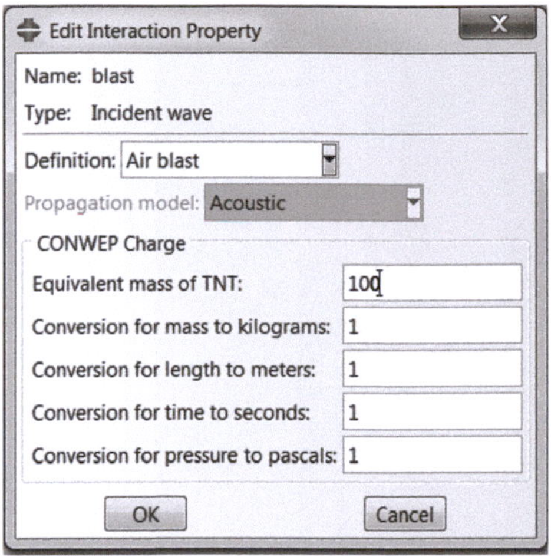

FIGURE 6.17 Defining the parameters for blast loading. (Abaqus® screenshot reprinted with permission from Dassault Systèmes.)

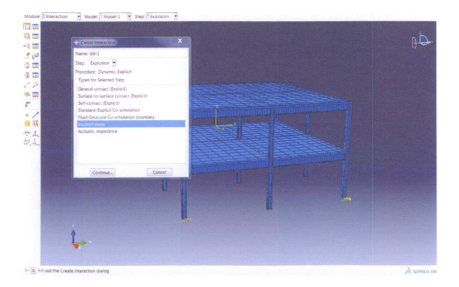

FIGURE 6.18 Define the location of the blast loading. (Abaqus® screenshot reprinted with permission from Dassault Systèmes.)

- It will then require the surface the blast wave will be acting on, as shown in Figure 6.19. The surfaces are chosen.
- Click on *Done*. Another window will pop up (Figure 6.20). Choose the *Wave Property Blast* (which we have defined) and input the required CONWEP data, such as time of detonation and magnitude scale factor.

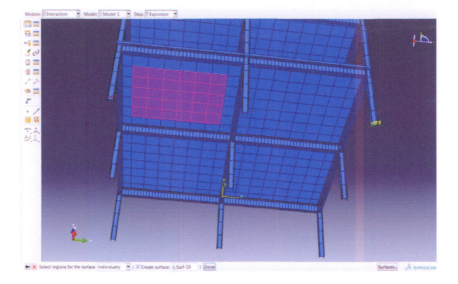

FIGURE 6.19 Selecting the surface. (Abaqus® screenshot reprinted with permission from Dassault Systèmes.)

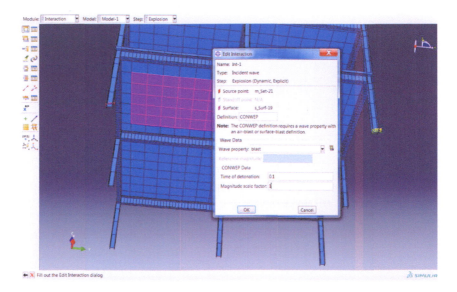

FIGURE 6.20 Defining the blast parameters. (Abaqus® screenshot reprinted with permission from Dassault Systèmes.)

6.6.4 Modelling Result

The model is analysed and the results are shown in the following sections.

6.6.4.1 Contour Plots The stress distribution and the acceleration distribution of the structure can be plotted in the ODB files of Abaqus® (Figures 6.21 and 6.22).

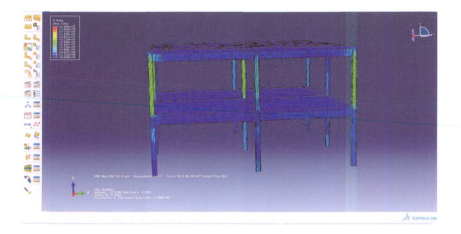

FIGURE 6.21 Stress distribution after blast loading. (Abaqus® screenshot reprinted with permission from Dassault Systèmes.)

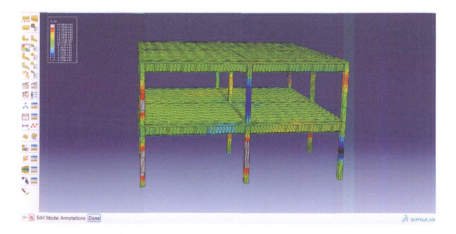

FIGURE 6.22 Horizontal acceleration distribution after blast loading. (Abaqus® screenshot reprinted with permission from Dassault Systèmes.)

6.6.4.2 *Time History of Certain Parameters*

- Click on *XY Data*. A window will pop up (Figure 6.23).
- Choose *ODB History Output* (Figure 6.24).

The time histories of the parameters, such as the external energy, can be extracted as shown in Figures 6.25 through 6.28. From the figures, readers can evaluate the response of the building under blast loading and perform the design of the building. However, for

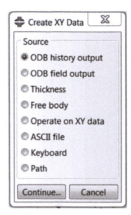

FIGURE 6.23 Selecting history output. (Abaqus® screenshot reprinted with permission from Dassault Systèmes.)

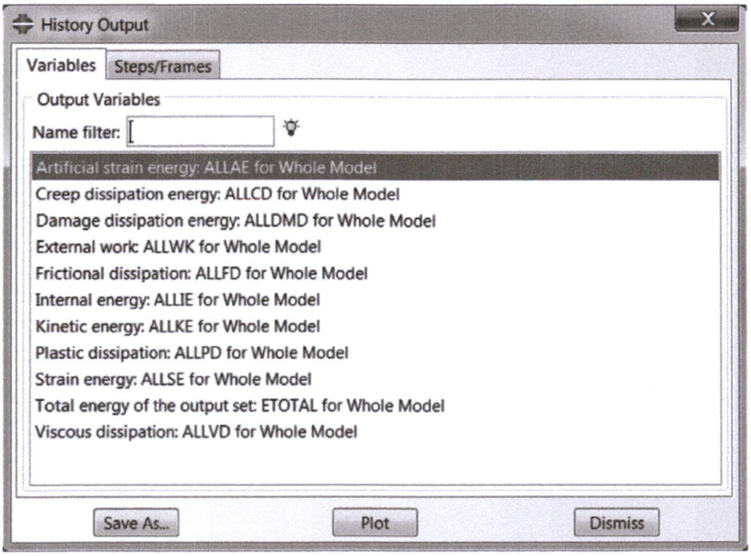

FIGURE 6.24 Selecting energy output. (Abaqus® screenshot reprinted with permission from Dassault Systèmes.)

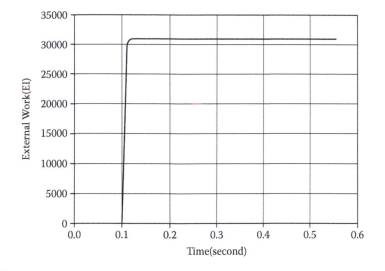

FIGURE 6.25 Time history of external work of whole model.

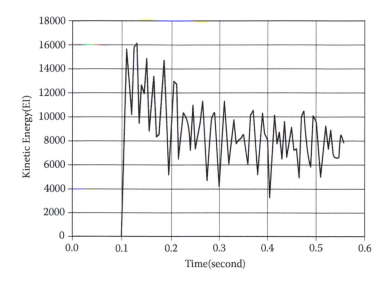

FIGURE 6.26 Time history of kinetic energy of whole model.

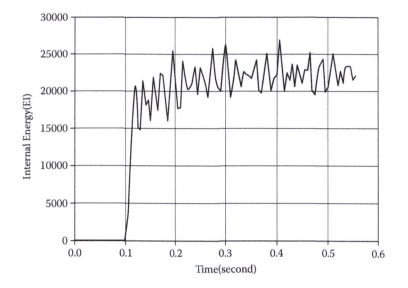

FIGURE 6.27 Time history of internal energy of whole model.

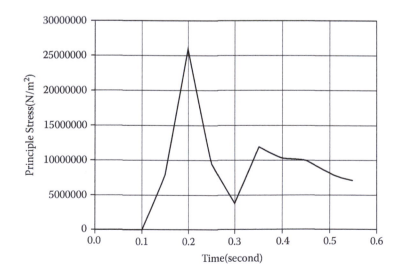

FIGURE 6.28 Principal stress of slab elements near the detonation points.

multistorey buildings, 3D solid elements introduced in this case study will bring dramatic computational cost, and it is not practical to build a multistorey building with all structural members represented using solid elements; therefore, the method introduced in Section 6.7 is more practical.

6.7 Modelling Examples of Progressive Collapse Analysis of Alfred P. Murrah Federal Building under Blast Load Using Abaqus®

6.7.1 Introduction

In this section, the blast analysis for the Murrah Federal Building is demonstrated using Abaqus®. For a detailed modelling technique, refer to Fu (2013). Similar to Chapter 5, the building is first built in Abaqus® to perform the blast analysis to identify the failure members. In the second step, using a procedure similar to that introduced in Chapter 2, a member removal analysis can be performed to determine the progressive collapse potential of the structure. As the purpose of this modelling example is to demonstrate the way to model the global behaviour of the Murrah Federal Building, some reasonable simplifications were made in the analysis.

6.7.2 Prototype Building

The Alfred P. Murrah Federal Building was modelled with Abaqus®. The layout of the building is shown in Figure 6.2. Due to the large number of elements, it was not practical to set up a 3D solid model using the same method introduced in Section 6.6. The model was first set up using the 3D modelling program ETABS (Figure 6.29). Then, using the program developed by Fu (2009), the model was converted into INP files for Abaqus®.

6.7.3 Applying Blast Load

In the blast analysis, a general-purpose program, ATBLAST (Applied Research Associates, 2000), was used for predicting explosive effects. ATBLAST is commercial software for evaluating potential blast damage. It is designed based on the empirical formula of UFC 3-340-02 (UFC, 2008). It calculates the blast loading parameters from an open hemispherical explosion based on the distance from the device. The program allows the user to enter the weight of the explosive charge, a reflection angle, the minimum and maximum ranges

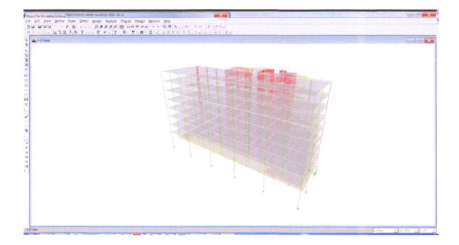

FIGURE 6.29 Alfred P. Murrah Federal Building setup in ETABS. (ETABS screenshot reprinted with the permission of Computer and Structures.)

to the charges, and the calculation interval. From this information, it can calculate the shock velocity, time of arrival, overpressure, impulse, and load duration of the blast loading. Using a program developed by Fu (2013), the blast loading profile by ATBLAST can be worked out as correspondent amplitudes that can be applied directly into the Abaqus® model.

In the analysis model, the blast was detonated in the same location of the lorry as it is introduced in Section 6.2.1, which is close to column G20, with an equivalent of 1800 kg of TNT applied. The blast profile (Figure 6.30) was extracted from the Abaqus® INP file and applied to the building.

6.7.4 Modelling Techniques

All the beams and columns were simulated using *BEAM elements. The orientation of a beam cross section is defined in Abaqus® in terms of a local, right-hand (t, n_1, n_2) axis system, where t is the tangent to the axis of the element, positive in the direction from the first to the second node of the element. n_1 and n_2 are basis vectors that define the local 1- and 2-directions of the cross section. n_1 is referred to as the first beam section axis, and n_2 is referred to as the normal to the beam (Figure 6.31). In the developed program by Fu (2013), the blast pressure worked out will be projected in the n_1 and n_2 directions on each beam and column.

The slabs and wall were simulated using the four-node *SHELL elements. Reinforcements were represented as a smeared layer in each shell element using the *REBAR elements and were defined in

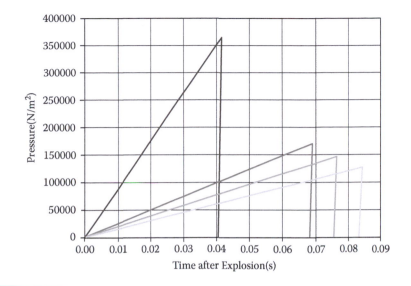

FIGURE 6.30 Blast profile.

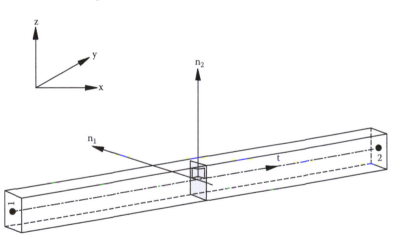

FIGURE 6.31 Local axis definition for beam-type elements in Abaqus®. (Reproduced with permission from Dassault Systèmes.)

both slab directions. In Abaqus®, the local material 1- and 2-directions lie in the plane of the shell. The default local 1-direction is the projection of the global 1-axis onto the shell surface. If the global 1-axis is normal to the shell surface, the local 1-direction is the projection of the global 3-axis onto the shell surface. The local 2-direction is perpendicular to the local 1-direction in the surface of the shell, so that the local 1-direction, local 2-direction, and positive normal to the surface form a right-hand set (Figure 6.32). In the developed

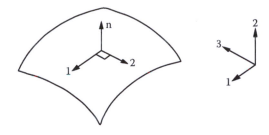

FIGURE 6.32 Default local shell material directions in Abaqus®. (Reproduced with permission from Dassault Systèmes.)

program by Fu (2013), the blast pressure worked out is projected in the local 3-direction on the shell panel.

The beam and shell elements were then coupled together using rigid beam constraint equations to give the composite action between the beam elements and concrete slabs. The concrete was modelled using a concrete damage plasticity model. The material properties of all the structural steel components and slab reinforcement were modelled using an elastic–plastic material model incorporating the material nonlinearity. The model is supported at the bottom (see Figure 6.34). The mesh representing the model was studied and is sufficiently fine in the areas of interest to ensure that the developed forces can be accurately determined.

6.7.5 Major Abaqus® Commands Used in the Simulation

The INP file of Abaqus® consists of several main parts. Readers can refer to the Abaqus® manual for detailed examples. Here only the most important parts are explained in detail, especially the amplitude of the blast profile, defined in Part 7, and the blast analysis step, shown in Part 9.

1. Coordinates (define the coordinates of all the nodes)

```
*node,nset = Node
1,25.13016,16.36888,4
2,37.30223,16.22658,4
3,49.96053,16.22658,4
4,12.66524,19.56854,4
5,12.66524,16.36888,4
6,6.511823,.04942671,0
. . . . . .
```

2. Frame element (define concrete beams and columns)

```
*element, type = b31,elset = COLUMN-CON1
1,746,839
```

```
*beam section, section = R,
elset = COLUMN-CON1,material = Con-beam
1.2,1.2
0,1,0
*element, type = b31,elset = COLUMN-CON2
2,747,840
*beam section, section = R,
elset = COLUMN-CON2,material = Con-beam
1.2,1.2
0,1,0
. . . . . . . . .
```

3. Shell element (define the walls and slabs)

```
*element, type = s4r,elset = WALL1
100001,766,859,854,761
. . . . .
*element, type = s4r,elset = DECK1
100039,877,915,914,910
. . . . . .
```

4. Section properties (define the shell element for concrete walls and slabs)

```
*shell section,elset = wall1,material = Concrete
0.45,9
*rebar layer (here the reinforcement in the shell
  elements are defined)
a252x,50.26e-6,0.200,0.03,s460,,1
a252y,50.26e-6,0.200,0.03,s460,,2
*shell section,elset = DECK1,material = Concrete
0.4,9
*rebar layer
a252x,50.26e-6,0.200,0.03,s460,,1
a252y,50.26e-6,0.200,0.03,s460,,2
```

5. Material

```
*material,name = C1 (concrete material is defined
  here)
*Concrete
4e+07, 0.
4.5e+07, 0.0025
*Failure Ratios
1.16, 0.056, 1.28, 0.33
*Tension Stiffening
1., 0.
0.01, 0.0025
```

```
*Density
2400.,
*Elastic
3.8e+10, 0.2
*material, name = s460 (steel rebar material will
  be defined)
*elastic
205000e6,0.3
```

6. Support and boundary conditions

```
*nset,nset = bottom node
939
934
. . . . .
*boundary
bottomnode,1,6
```

7. Amplitude of blast profile

```
*amplitude,name = A100001 (blast profile for shell
elements, wall and slab)
0,0,.08307,0,.08407,127140.112,.10963,0
*amplitude,name = A100002
0,0,.07548,0,.07648,145273.436,.10104,0
. . . . . . .
```

8. Analysis step 1 (static step)

```
*Step, name = Static
*Static
0.25, 1., 1e-05, 1.
*Dload (Define live load)
DECK1,p,-0.0625e3
all, GRAV, 9.81, 0., 0., -1. (Define Gravity load)
*Restart, write, frequency = 1 (OUTPUT REQUESTS)
*Output, field, variable = PRESELECT (Define FIELD
  OUTPUT: F-Output-1)
*Output, history, variable = PRESELECT (Define
  HISTORY OUTPUT: H-Output-1)
. . . . . . (please refer to Chapter 2 for detailed
  explanations)
*End Step
```

9. Analysis step 2 (blast profiles are applied to the beams, columns, walls, and slabs)

```
*step,inc = 10000
*dynamic,haftol = 80000000,initial = no
```

```
0.00025,0.4,0.0000001,0.0005
*Dload,amplitude = A100001 (defining area blast
  load on shell elements)
100001,P,1 (applied in the normal direction, with
  the scale factor 1)
. . . . . . .
*Dload,amplitude = A1 (defining line load on beam
  elements, applied in the normal)
1,P1,1 (applied in the local 1 direction of the
  beam element with the scale factor 1)
. . . . . . . .
*Dload,amplitude = B1 (defining line load on beam
  elements)
1,P2,1 (applied in the local 2 direction of the
  beam element, with the scale factor 1)
```

6.7.6 Modelling Result

The model is analyzed and the results are shown below.

6.7.6.1 Contour Plots

- Go to *Results* menu and click on *Field Output*. A window will pop up (Figure 6.33).
- Choose *PDLOAD* (pressure from distributed loads on element face) (Figure 6.33). The blast pressure will be shown (Figure 6.34). It can be seen that the blast pressure is propagating from the first floor to the top levels.

Similarly, the stress contour can also be checked, as shown in Figure 6.35.

6.7.6.2 Time History of Certain Parameters
As the blast was detonated close to column G20 (refer to Figure 6.2 for the location of column G20), the axial and shear forces inside column G20 were checked. As mentioned in Section 6.2.1, the adjacent two columns, G16 and G24, were also destroyed by the large shear force produced by the blast wave; therefore, the shear force inside these two columns was also checked.

It can be seen that there were huge axial and shear forces observed in column G20, and this column was destroyed (Figures 6.36 and 6.37).

From Figures 6.38 and 6.39, it can be seen that there was also a huge shear force observed in columns G16 and G24. This column was destroyed by the blast as well. However, as it was located far

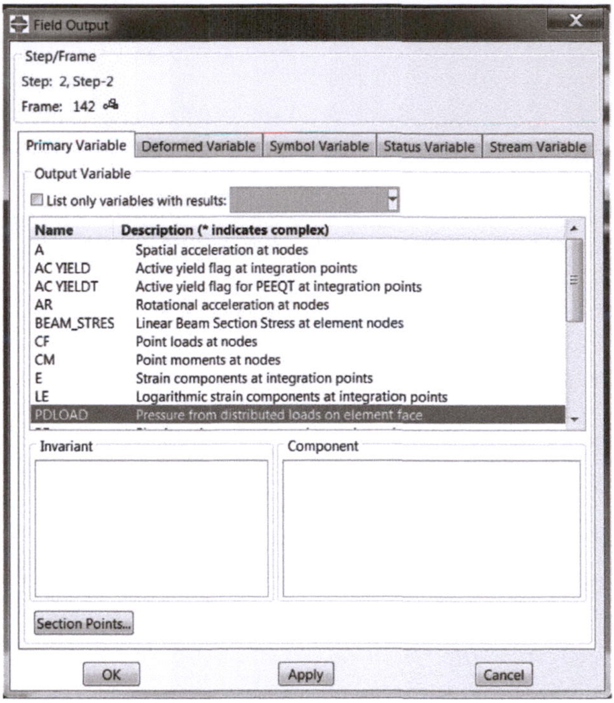

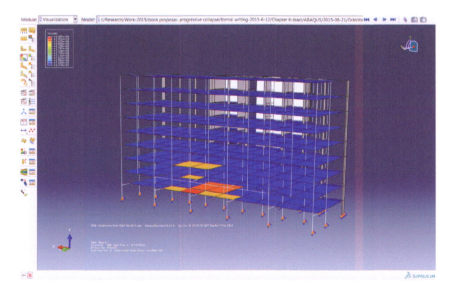

FIGURE 6.33 Choosing PFLOAD. (Abaqus® screenshot reprinted with permission from Dassault Systèmes.)

FIGURE 6.34 Blast loading pressure distribution. (Abaqus® screenshot reprinted with permission from Dassault Systèmes.)

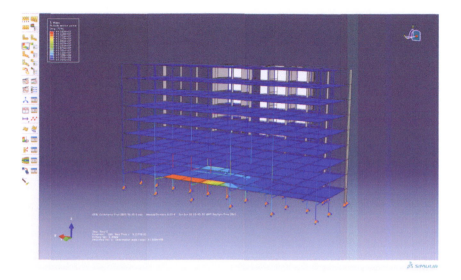

FIGURE 6.35 Stress distribution after blast loading. (Abaqus® screenshot reprinted with permission from Dassault Systèmes.)

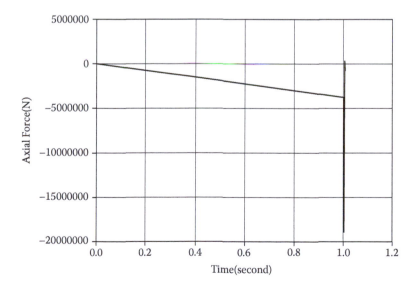

FIGURE 6.36 Axial force of G20 after blast.

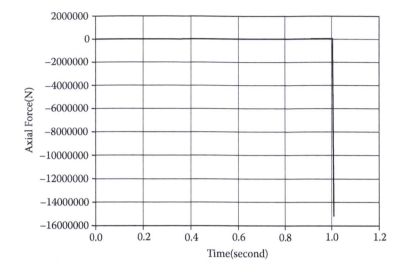

FIGURE 6.37 Shear force of G20 after blast.

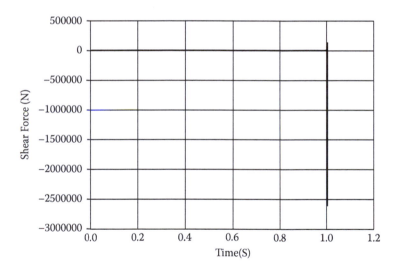

FIGURE 6.38 Shear force of G16 after blast.

from the detonation location, the shear force was dramatically reduced compared to that for column G20 (Figure 6.3).

6.7.7 Progressive Collapse Potential Check

After the investigation in Section 6.7.6, it is noticed that columns G20, G16, and G24 were all destroyed due to the blast loading. Therefore, in this section, a column removal analysis, as demonstrated in Chapter 2, could be performed. In the analysis, columns

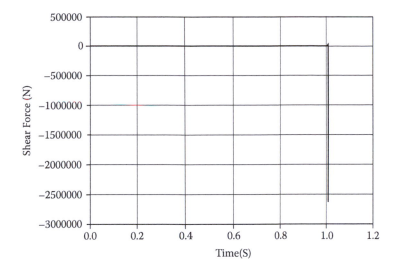

FIGURE 6.39 Shear force of G24 after blast.

G20, G16, and G24 would be removed simultaneously. The internal force, such as the bending moment of the transfer beam, could be checked, and the progressive collapse potential could be assessed. In this case, the partial building collapse is quite obvious; therefore, the analysis was not performed.

References

Applied Research Associates. 2000. ATBLAST 2.0. Albuquerque, NM: Applied Research Associates.

Baker, W.E., Cox, P.A., Westine, P.S., Kulesz, J.J., and Strehlow, R.A. 1983. *Explosion Hazards and Evaluation*. New York: Elsevier Scientific.

BSI (British Standards Institution). 2005. Scheduling, dimensioning, bending and cutting of steel reinforcement for concrete—Specification. BS 8666. London: BSI.

Cormie, D., Mays, G., and Smith, P. 2009. *Blast Effects on Buildings*. 2nd ed. London: Thomas Telford.

Cranz, C. 1926. *Lehrbuch der Ballistik*. Berlin: Springer.

FEMA (Federal Emergency Management Agency). 2003. Reference manual to mitigate potential terrorist attacks against buildings. Risk Management Series, FEMA 427. Washington, DC: FEMA, December.

Fu, F. 2009. Progressive collapse analysis of high-rise building with 3-D finite element modeling method. *Journal of Constructional Steel Research*, 65(6), 1269–1278.

Fu, F. 2013. Dynamic response and robustness of tall buildings under blast loading. *Journal of Constructional Steel Research*, 80, 299–307.

Hetherington, J., and Smith, P. 1994. *Blast and Ballistic Loading of Structures.* Boca Raton, FL: CRC Press.

Hopkinson, B. 1915. British ordnance board minutes 13565.

Jarrett, D.D. 1968. Derivation of British explosives safety distances. *Annals of the New York Academy of Sciences,* 152, 18–35.

ODCEM (Oklahoma Department of Civil Emergency Management). 1995. After action report, Alfred P. Murrah Federal Building bombing, 19 April 1995 in Oklahoma City, Oklahoma. Oklahoma City: ODCEM.

Unified Facilities Criteria (UFC). 2008. Structures to resist the effects of accidental explosions, 2008 change 2. UFC 3-340-02. Washington, DC: National Technical Information Service.

Yandzio, E., and Gough, M. 1999. Protection of buildings against explosions. SCI Publication 244. Berkshire, UK: Steel Construction Institute.

CHAPTER 7

Conclusion

7.1 Introduction

In this book, design methods for preventing disproportionate collapse for different types of structures, such as multistorey buildings, space structures, and bridges, were discussed. Different loading regimes that can trigger the collapse of structures, such as fire and blast, were introduced. The collapse mechanisms of different types of structures were also analyzed. In addition, progressive collapse analysis methods were introduced and demonstrated using commercial programs through modelling examples of the Twin Towers, World Trade Center 7, Murrah Federal Building, and Millau Viaduct.

In this chapter, the relevant design and analysis methods are summarized.

7.2 Summary of Design Guidances and Methods

In this book, we introduced several design guidances for preventing disproportionate collapse. They are mainly for building structures, such as the Building Regulations 2010 (HM Government, 2013) and BS 5950 (BSI, 2001) in the UK, Eurocode EN 1990 (BSI, 2010) in Europe, CSA-S850-12(CSA, 2012) in Canada, and the Department of Defense (DOD, 2009), General Services Administration (GSA, 2003), ASCE Standard 7 (ASCE, 2005), and NIST (2007) in the United States. There are no major design guidances available regarding the disproportionate collapse of space structures. For bridge structures, PTI (2007) and FIB (2005) are two guidances with special requirements to make sure progressive collapse is not triggered.

Several design methods have been proposed by design guidances such as DOD (2009), GSA (2003), and BS 5950 (BSI, 2001) for building structures. They are divided into two major categories, direct design method and indirect design method, which include the design of a key element, the tying force method, and the alternative load path method.

7.3 Summary of the Analysis Method

For building structures, there are four basic analysis methods proposed by GSA (2003): linear static, linear dynamic, nonlinear static, and nonlinear dynamic. There is no clear analysis method in design guidances for space structure and bridge structures; therefore, the aforementioned methods can be used to analyze space structures and bridge structures in the current design practice.

7.4 Summary of Collapse Mechanisms and Measures to Prevent Progressive Collapse

7.4.1 Multistorey Buildings

For building structures, catenary action can be utilised to resist the progressive collapse of a building. Providing sufficient ties in both steel and concrete buildings, increasing the ductility, and providing alternative load paths are the most effective methods in progressive collapse design.

In addition, increasing the redundancy of a structure system will definitely enhance its resistance to progressive collapse. Some researchers have also developed retrofit methods, such as enhancing beam-to-column connections, using steel cables, or providing a backup system. However, these methods increase the cost of projects. Engineers should make selections based on the category of the building and the requirements of the client.

7.4.2 Long-Span Space Structures

The loss of some critical members due to an excessive gravity load, such as snow, will cause the collapse of long-span structures such as double-layer grids. For single-layer space structures, such as domes, local snap-through of certain critical members can cause global buckling. Therefore, an extra margin of safety should be made for the structural members to prevent progressive collapse due to abnormal gravity loads.

In addition to the above considerations, the space frame roof structure requires consideration for support flexibility in the design.

7.4.3 Bridge Structures

Bridge structures or continuous beam bridges can be designed to be span independent; therefore, the failure of one span will not trigger the collapse of the whole structure. Pier failure is one of the major reasons for bridge collapse. A pier protection method, such as an artificial island, can be used. Hanger failure is a major reason for the collapse of suspension bridges. Therefore, in the design of hangers, a large margin of safety should be used. Cable-stayed bridges feature high redundancy; however, they should be able to accommodate cable failure. In their design, checks should be made by removing one or several cables to determine the robustness of the bridges.

7.5 Conclusion

The main purpose of this book was to provide some guidance and case studies for engineers to perform design and analysis to prevent disproportionate collapse. This book is based on the best knowledge of the author. I sincerely hope readers get some benefit from this book.

References

ASCE (American Society of Civil Engineers). 2005. Minimum design loads for buildings and other structures. SEI/ASCE 7-05. Washington, DC: American Society of Civil Engineers.

BSI (British Standards Institution). 2001. Structural use of steelwork in buildings. Part 1: Code of practice for design—rolled and welded sections. BS 5950. London: BSI.

BSI (British Standards Institution). 2010. Eurocode—Basis of structural design: Incorporating corrigenda December 2008 and April 2010. BS EN 1990: 2002 + A1: 2005. London: BSI.

CSA (Canadian Standards Association). 2012. Design and assessment of buildings subjected to blast loads. CSA-S850-12. Ontario: CSA.

DOD (Department of Defense). 2009. Design of buildings to resist progressive collapse. UFC 4-023-03. Arlington, VA: Department of Defense, July 14.

FIB (International Federation for Structural Concrete). 2005. Acceptance of stay cable systems using prestressing steels. Lausanne: FIB.

GSA (General Services Administration). 2003. Progressive collapse analysis and design guidelines for new federal office buildings and major modernization projects. Washington, DC: GSA.

HM (Her Majesty's) Government. 2013. The Building Regulations 2010: Structure, A3: Disproportionate collapse. Approved Document A, 2004 edition, incorporating 2004, 2010, and 2013 amendments. London: HM Government.

NIST (National Institute of Standards and Technology). 2007. Best practices for reducing the potential for progressive collapse in buildings. Gaithersburg, MD: NIST, Technology Administration, U.S. Department of Commerce.

PTI (Post-Tensioning Institute). 2007. *Recommendations for Stay Cable Design, Testing and Installation*. 5th ed. Phoenix, AZ: Cable-Stayed Bridge Committee.

Index

Y

Young's modulus, 61

Z

zone model, 112

An environmentally friendly book printed and bound in England by www.printondemand-worldwide.com